Wiraputra S

MW00683321

Benchmarking Based Process Re-engineering for Construction Management

Wiraputra Sutan, Min-Yuan Cheng

Benchmarking Based Process Re-engineering for Construction Management

A Systematic Method to Learn From Best Practice Company

VDM Verlag Dr. Müller

Imprint

Bibliographic information by the German National Library: The German National Library lists this publication at the German National Bibliography; detailed bibliographic information is available on the Internet at http://dnb.d-nb.de.

Any brand names and product names mentioned in this book are subject to trademark, brand or patent protection and are trademarks or registered trademarks of their respective holders. The use of brand names, product names, common names, trade names, product descriptions etc. even without a particular marking in this works is in no way to be construed to mean that such names may be regarded as unrestricted in respect of trademark and brand protection legislation and could thus be used by anyone.

Cover image: www.purestockx.com

Publisher:
VDM Verlag Dr. Müller Aktiengesellschaft & Co. KG , Dudweiler Landstr. 125 a, 66123 Saarbrücken, Germany,
Phone +49 681 9100-698, Fax +49 681 9100-988,
Email: info@vdm-verlag.de

Zugl.: Taipei, National Taiwan University of Science and Technology, Diss., 2007

Produced in USA and UK by:
Lightning Source Inc., La Vergne, Tennessee, USA
Lightning Source UK Ltd., Milton Keynes, UK
BookSurge LLC, 5341 Dorchester Road, Suite 16, North Charleston, SC 29418, USA

ISBN: 978-3-639-00451-9

ABSTRACT

Business Process Reengineering (BPR) is considered to be the fastest methodology to be used in order to improve business process in a current fast-changing environment. BPR attempts to achieve dramatic improvements in critical measures of performance by using the power of modern information technology (IT) fundamentally to rethink and radically to redesign the business process. However, many construction industries proposed BPR to evaluate the business process, not to redesign the business process. This study focuses on the idea of "Business Process Reengineering (BPR)" in applying benchmarking philosophy to redesign the business process.

The construction industry has been slow to adopt competitive benchmarking as a tool for continuous improvement. Identifying, and adapting the best-practice process are the important factors of benchmarking. In order to determine the most suitable process from best practice companies' to be adapted in the benchmarking company, an accurate analysis of the gaps between best-practice processes and benchmarking company's process is essential for the redesign of business processes.

I

This research integrates BPR philosophy, semantic similarities concept and trend model concept to develop a benchmarking-oriented process reengineering (BOPR) that enables a project team to determine the most suitable process from the best practice company. A four-phased process analysis including business process modeling, process similarities analysis, process communication index analysis and process adaptability calculation was developed. An approach of applying the concepts of semantic similarities analysis to find the semantic-related objects between best-practice processes and benchmarking process was also proposed. In addition, trend model concept was applied to evaluate the degree of communication ease of best-practice processes when it is adapted in a benchmarking company.

Summarily, by referring the proposed BOPR method, project team will have better understanding in determining the most suitable process from best-practice companies to be adapted in benchmarking company.

Keywords: **benchmarking, business process reengineering (BPR), semantic similarity.**

ACKNOWLEDGEMENTS

Thank you to God for the strength and guidance that lead me to finish this dissertation. These acknowledgements attempt to thank people who supported and helped me make this dissertation a reality.

First, I wish to express my sincere gratitude to my advisor, Prof. Min-Yuan Cheng, for his helpful suggestions, professional advice, and patience in guiding me to the right research directions. Thanks are also extended to Dr. Ming-Hsiu Tsai, for serving on my supervisory committee, and assistance for conducting parts of this research. I am also indebted to Energy Lian, Dennie Chen, Hsing Chih and Yu Wei for kind assistance in helping me to contact and liaise with Construction Companies.

Thanks are also due to doctoral degree students and master degree students: Roy, Amber, James, Matt, Ivan, Andy, Erick, Jian Chong, Jia Liang and Wu Xiang in helping me to do the communication resistance survey.

I am extremely grateful to Evangelist Jeffrey for his time, patience and continuous encouragement in providing me editorial comments. Furthermore, I would like to thank Steven Khoe, Arief, Jerry and Reymont for sharing my happiness and sadness during this stressful portion of my life. Special appreciation is also extended to BAI team for the friendship that makes me realize I'm not a stranger in this world.

Last, but certainty not least, I would like to thank my family that give me endless support and encouragement during these years of my studies, for which I will always be grateful. Soli deo Gloria.

Taipei, Taiwan   Wiraputra Sutan (陳世瑋)

July 2007

TABLE OF CONTENTS

ABBREVIATIONS AND SYMBOLS

Abbreviations

AI	Adaptability Index
ARIS	Architecture of Integrated Information Systems
ARM	Activity Relationship Matrix
ASim	Activity Similarity
BOPR	Benchmarking Oriented Process Reengineering
BPM	Business Process Modeling
CR	Consistency Ratio
CRA	Communication Resistance Analysis
CRM	Communication Resistance Matrix
PAA	Process Adaptability Analysis
PISim	Process Information Similarity
PFSim	Process Functional Similarity
PSA	Process Similarity Analysis
SS	Semantic Similarity
TCI	Total Communication Index
TRI	Total Resistance Index

Symbols

a	explicit affinity associated with the relationship r_{ij}
a_{hj}	explicit affinity of n_h and n_j
a_{ih}	explicit affinity of n_i and n_h
a_{ij}	explicit affinity of relationship of n_i and n_j
$a_{ij\text{-}R1}$	explicit affinity of R_1 relationship of n_i and n_j
$a_{ij\text{-}R2}$	explicit affinity of R_2 relationship of n_i and n_j
a_{BT}/a_{NT}	explicit relationship affinity of BT/NT
a_{RT}	explicit relationship affinity of RT
a_{SYN}	explicit relationship affinity of SYN
A_{het}	heterogeneous semantic affinity
A_{hom}	homogeneous semantic affinity
A_{ih}	name of hth activity of the *process i*
A_{jk}	name of kth activity of the *process j*
A_{imp}	affinity function of the implicit relationship
$AC(n_i, n_j)$	affinity coefficient of two names n_i and n_j
AIN_{ih}	input set of A_{hi}
$AOUT_{jk}$	output set of A_{jk}
$ASim()$	activity similarity function of two analyzed activities
$A(X,Y)$	name set affinity parameter of the X and Y name sets
BT/NT	Broader/Narrow term semantic relationship
CL_{ij}	common concept link of n_i and n_j in the data semantic hierarchy
f	specific name set involved in ζ
$f1$	process name of the process textual model
$f2$	process input data set of the process textual model

$f3$	process output data set of the process textual model
$f4$	activity set of the process textual model
k	number of occurrences of the relationship type for the considered pair of names
Ki	communication resistance value
l_A	hierarchy length in the activity semantic hierarchy
l_{ij}	semantic linkage of n_i and n_j
l_{ij}	hierarchical length of the common link of n_i and n_j
l_D	hierarchy length in the data semantic hierarchy
L_{CL-ij}	hierarchy level of the common concept link of n_i and n_j
n_j	name of the j data entity in the set j
n_r	name of output data entity r
n_s	name of input data entity s
$NA(n_i,n_j)$	name affinity of entities n_i and n_j
P_i	process i
P_i	process j
$PISim()$	process information similarity function
PI_{ij}	process information similarity of jth process in the company i
$PFsim()$	process functional similarity function
PF_{ij}	process functional similarity of jth process in the company i
r_{ij}	type of semantic relationship in a semantic linkage
RT	Related term semantic relationship
SS_{ij}	semantic similarity of n_i and n_j
SYN	synonymy semantic relationship
TC_{ij}	total communication index of jth process in the company i
$|P_i|$	number of activities of process i
$|P_j|$	number of activities of process j

W_1	weight of process information similarity
W_2	weight of process functional similarity
W_3	weight of total communication resistance
δ	initial similarity value for the bottom level concept links
ζ	name sets derived from the process textual model
$\mathfrak{R}$	explicit relationship
$\mathfrak{R}_{\text{hom}}^{k}$	homogeneous multiple explicit relationship
$\Rightarrow$	implicit relationship
$\Rightarrow_{\text{hom}}^{k}$	homogeneous multiple explicit relationship

LIST OF FIGURES

LIST OF TABLES

Chapter 1

INTRODUCTION

1.1 Research Motivation

The phrase "business process reengineering" (BPR) first appeared in 1990, raised by Michael Hammer in a Harvard Business Review paper called "Reengineering work: Don't automate, obliterate" (Hammer, 1993). Hammer believed that, although in the fiercely competitive environment of the 1990s most businesses were adopting measures such as rationalization and automation to improve their organizations, none of these measures was truly improving business operations. To solve these problems, the idea of "reengineering" was advanced as a theory and tool for business reorganization. In Hammer's fundamental definition, BPR starts from very basic issues and asserts that the reengineering process can dramatically improve an organization in terms of its costs, quality, services and speed. For this purpose, three BPR "cores" are insisted on in Hammer's article: process reorganization, the use of information technology, and organizational redesign (Hammer and Champy, 1993); that is, a successful implementation of BPR depends not only on business process mechanism, but organization structure should also be considered.

As is apparent from the above, BPR attempts to achieve dramatic improvements in critical measures of performance by using the power of modern information technology (IT) fundamentally to rethink and radically to redesign the business process (Hammer and Champy, 1993). However, in most researches, BPR was aimed to evaluate the business process performance, not to redesign the business process (Cheng et al., 2003). To solve this problem, a reliable methodology to

1

redesign the business process in accordance to BPR is essential and significant for construction companies.

In BPR, to redesign a process by using trial and error methodology may be time consuming due to lack of information and experiences. Furthermore, there is no guarantee of the new redesigned process' performance because the new redesigned process' performance can only be validated it by implementing it. Thus, BPR is usually being treated as a high risk solution to enhance the business performance. Therefore, if the company can redesign its process based on best-practice companies' process, then the duration of BPR might be shortened. The best-practice process implies the high performance process that will decrease the risk of implementing BPR and increase the success probability of BPR. Based on this idea, this study applies benchmarking philosophy in accordance to business process reengineering (BPR) to redesign the business process.

Benchmarking originated in the mid-1970s and is a process of continuously comparing and measuring an organization with leaders anywhere in the world to gain information that will help it take action to improve its performance (Camp, 1995). Benchmarking is an excellent source of business ideas. One of the primary benefits of process benchmarking is that it exposes individuals to new products, work process and ways of managing company resources (Spendolini, 1992). Moreover, benchmarking offers an easier way to redesign process by providing a path or blue print of targeted process from best-practice companies and serves to motivate project team by establishing realistic goals demonstrated to be achievable from best-practice companies. Thus, following to BPR philosophy, this study proposes benchmarking philosophy to identify best-practice process and to intrigue BPR for adapting the best-practice process in benchmarking company.

Learning from best-practice companies is the key point of process benchmarking but the main problem is to determine which best-practice companies is the best-to-learn for the benchmarking company. However, most researches in construction industry focused on discovering the best-practice companies from the industry and only few focused on how to determine the most-suitable companies to be adapted among several selected best-practice companies (Lee et al. 2005; Ramirez et al. 2004; O'Connor and Miller. 1994). Basically, in process benchmarking, after several companies are discovered to be the best-practice companies in the industry that can be learned by benchmarking company, then a project team must determine which best-practice companies is the most suitable company to be learned from. Since this study focuses on redesigning a process, benchmarking philosophy is applied to provide the best-practice processes from best-practice companies to be paradigm in redesigning a process. However, to determine the most suitable process from best-practice companies for benchmarking company is the main problem to be solved. Logically, the most suitable process to be adapted in the benchmarking company is the process which has similar characteristics with benchmarking process and can be executed smoothly when it will be performed in benchmarking company. Therefore, the similarity between best-practice processes and benchmarking process characteristics is necessary to be evaluated to confirm the corresponding relationship between two processes. Moreover, since the selected best-practice process will be performed in the benchmarking company, the execution resistance of best-practice processes when it is performed in benchmarking organization needs to be calculated. To overcome these problems, the semantic similarities concept is applied to evaluate the similarity of data and activities between best-practice processes and benchmarking process. Meanwhile, the trend model concept is applied to evaluate the degree of communication ease from the best practice processes when it is performed in the

3

benchmarking company.

Summarily, in order to assist project team to redesign a process based on the most suitable best-practice process, this research integrates BPR philosophy, semantic similarities concept and trend model concept to develop a benchmarking oriented process reengineering (BOPR) method to determine the most suitable process from the best-practice companies. A four-phased process analysis including business process modeling, process similarities analysis, process communication index analysis and process adaptability calculation was developed in achieving the goal.

1.2 Research Objectives

The primary purpose of this study is to develop a systematic analysis method to assist project teams in determining the most suitable process from best-practice companies to be performed in benchmarking company's. The sub-objectives required to achieve the primary goal are the followings:

◆ Following the business process reengineering philosophy in applying benchmarking philosophy to identify the performance of best practice process and to intrigue the process to be adopted.

◆ Developing process similarity analysis to confirm the corresponding semantic relationships of each compared data object pair and function object pair between best-practice processes and benchmarking process.

◆ Developing process communication index analysis to evaluate the degree of communication ease of best practice processes to be performed in benchmarking company organization.

◆ Developing process adaptability calculation to create an Adaptability Index (AI) that represents the acceptance degrees of each best practice company's process to be adapted in benchmarking company.

4

1.3 Scope Definition

1.3.1 Boundary Identification

Benchmarking oriented process reengineering method is significant to facilitate the project team in determining best practice process in benchmarking company's process. The boundary of this research is specified below:

◆ This thesis focuses on identifying and determining the most suitable procurement process from best practice companies in construction industries.

◆ This research aims at analyzing a procurement process based on data, function and organization views of process in order to facilitate project team to determine the most suitable procurement process for benchmarking company.

1.3.2 Research Assumptions

To facilitate project team in identifying and determining the most suitable procurement process from best practice companies, three assumptions need to be explored before the methods could be developed in this study,

◆ The data information and process information regarding procurement process from best-practice companies and benchmarking company have been identified.

◆ The most suitable process is the process which has similar characteristics with benchmarking process and can be executed smoothly when it will be performed in benchmarking company.

◆ Since the company cultures are too complicated to be quantified, these factors are beyond the scope of this study.

1.4 Research Methodology

The methodology used to achieve the research objectives is summarized in Figure 1.1 and 1.2. Figure 1.1 depicts the flow chart of the general research process while Figure 1.2 shows the detailed flow chart of the general flow chart in detail. Referring to Figure 1.2, the blocks on the left side are the procedures used to accomplish this study. The blocks on the right side are detailed methods and attributes concerned with execution of the tasks on the left side.

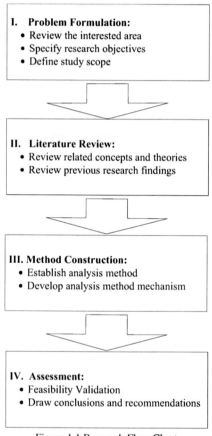

Figure 1.1 Research Flow Chart

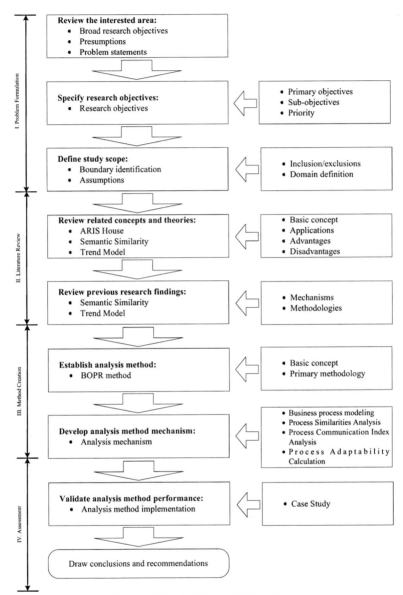

Figure 1.2 Detailed Research Flow Chart

1.4.1 Problem Formulation

Review the Interested Area

This study is interested in solving construction management problems. In the preliminary research, this study reviews the most difficult parts faced in benchmarking process for procurement process, analyzing the similarity between two processes, methods of benchmarking process analysis and applications of benchmarking process analysis methods. This approach relies on two main resources: interviews with experts and previous research works.

Specify Research Objectives

After reviewing the problem area, primary research objectives are defined to solve the addressed collaboration problems of benchmarking process. In order to systematically achieve the primary goals, sub-objectives are identified. In addition, the primary objectives and sub-objectives are prioritized to make study approach more logical and achievable.

Define Study Scope

The scope of a study should be achievable with a reasonable time and effort. Therefore, research boundary is distinguished. In this step, requisite assumptions are also established to reduce complexity of the process benchmarking problems.

1.4.2 Literature Review

Review Related Concepts and Theories

At this junction, the researcher undertakes extensive survey of concepts and theories connected to the formulated problem to broaden the knowledge base. Academic journals, conference proceedings, books, websites, published and unpublished reports, etc. are tapped depending on the problem.

In creating benchmarking process analysis method to facilitate project team, there are three concept and techniques needs to be reviewed, ARIS house concept and techniques to depict the data, function, and organization views, semantic similarity concept to find the semantic-related data objects and function object in different processes and trend model concept to evaluate the communication resistance of process. Consequently, basic concepts, applications, advantages, disadvantages of ARIS house, semantic similarity analysis, and trend model are reviewed.

Review Previous Research Findings

Previous research findings provide a basement for further research. Previous discoveries related to the objectives of this study are investigated through formal literatures including academic journals, conference proceedings, books, and published reports.

Benchmarking is considered to be one of the fastest methodologies to be used in implementing business process improvement in a current fast-changing environment. However, due to the lack of information, motivations, and experiences regarding benchmarking have caused some confusion on the true meaning of benchmarking and its relevance to business organizations in construction industries. Thus, this study investigates the method of benchmarking process analysis of procurement process in construction industries.

1.4.3 Method Creation

Establish Analysis Method

The purpose of this stage is to establish Benchmarking Oriented Process

Reengineering (BOPR) method which can facilitate project team in identifying, determining and adapting the best procurement process from best practice companies to be implemented in benchmarking company's procurement process. Thus, four phases, namely (1) business process modeling, (2) process similarity analysis, (3) process communication index analysis, and (4) process adaptability calculation, are necessary involved in the method.

Develop Analysis Method Mechanism

Based on the analysis method created in the previous step, a mechanism for each phase is elaborated. Moreover, the basic concepts and the primary methodologies related to the elements for each phase in BOPR method are determined.

1.4.4 Assessment

Feasibility Validation

The feasibility of each phase in BOPR method is validated through method of implementation. A case study is applied to address methods and mechanisms for each phase in BOPR. The feasibility of the method can be validated using the feedback from the case study.

Conclusions and Recommendations

This research is concluded from reviewing the study objectives and identifying the research contributions. The lessons learned are also discussed. As well, directions are made for further advancement.

1.5 Study Outline

This thesis consists of five additional chapters. Each chapter is briefly described.

Chapter 2 serves as an introduction to related methods that are used for achieving the study's objectives. Architecture of Integrated Information Systems (ARIS) house, semantic similarity and trend model are introduced for those who are not familiar with these methods. Basic concept, and applications related to each method are discussed in detail.

Chapter 3 describes the Benchmarking Oriented Process Reengineering (BOPR). The analysis method consists of four phases, namely (1) business process modeling, (2) process similarity analysis, (3) process communication index analysis, and (4) process adaptability calculation is proposed and briefly introduced in this chapter.

Chapter 4 presents the BOPR method to facilitate project team for determining the most suitable procurement process. Method concept, analysis phase and method adaptation from the case study are presented.

Chapter 5 is a review of the research purposes along with the summary, conclusions and recommendations. Research contributions and lessons learned are also stated. Finally, this chapter addresses prospects for future study.

Chapter 2

LITERATURE REVIEW

2.1 Benchmarking

2.1.1 Overview

Benchmarking is the search for the best practices that will lead to superior performance of an organization (Camp, 1989). It is a quality concept that has captured the interest of many businesses, and has been gaining popularity among executives and senior managers (Camp, 1989). The subject has triggered considerable interest although there is still some confusion on the true meaning of benchmarking, its relevance to business organizations, whether in manufacturing or service sectors, and how it could be successfully implemented are not yet fully understood (Zairi, 1992).

> **Definition.** Benchmarking is a systematic and continuous measurement process; a process continuously measuring and comparing organization's business process against business leaders anywhere in the world to gain information which will help organization to take action to improve its performance (Planning, 1992).

The benchmarking philosophy is proposed by the Japanese, who have used the term *dantotsu*, which means striving to be the best of the best (Camp, 1989). This business philosophy has been applied in Japan since the end of the World War II (Taiichi, 1990). Accordingly, the Western countries where the term benchmarking has been applied fairly successfully to a large number of business organizations adopted the benchmarking philosophy from the Japanese. Indeed, the first Western company to adopt benchmarking practice for its products and processes, Xerox, made its initial comparison with its Japanese affiliate Fuji-Xerox and later with other Japanese competitors in 1979 (Camp, 1989). The results were very revealing. Xerox established that its Japanese competitors were selling photocopy machines at what it costs Xerox to produce them.

Terms Used in Benchmarking

In the course of developing benchmarking as a formal management tool, associated terminologies have also emerged. Some of the key terminologies and their definitions are (Watson, 1993):

♦ *Benchmark*: a measured best-in-class achievement, a reference or measurement standard for comparison; a performance level recognized as the standard of excellence for a specific business practice.

♦ *Best practice*: superior performance within an activity, regardless of industry, leadership management, or operational approaches, or methods that lead to exceptional performance.

♦ *Enabler*: the process, practice, or methods that facilitate the implementation of a best practice and help meet a critical success factor; characteristics that help explain the reasons for the achievement of benchmark performance.

♦ *Entitlement*: the best can be achieved in the process performance using current resources to eliminate waste and improve cycle time; obvious improvements that are identified during benchmarking and may be accomplished as short-term goals.

♦ *Best in class*: outstanding process performance within an industry; synonymous term is *best of breed*.

◆   *Benchmarking gap*: a difference in performance, identified through a comparison, between the benchmark for a particular activity and other companies; the measured leadership advantage of the benchmark organization over others.

◆   *World class*: leading performance on a process, independent of industry, or geographic location, as recognized using process benchmarking for comparison to other world contenders.

Since benchmarking process is a continuous systematic process for evaluating the products, services and work processes of organizations that are recognized as representing best practices for the purpose of organizational improvement (Spendolini, 1994), a common understanding of benchmarking has to be reached. The remainder of this section will review the objectives of benchmarking, types of benchmarking and benchmarking process model.

2.1.2 Objectives of Benchmarking

Benchmarking is a positive, proactive process that can change business operations in a structured fashion to achieve superior performance. Benchmarking aims at ensuring that the best practices are followed in an ever-changing environment. The process provides a management tool for measuring and comparing any part of an organization's operation, product, or service against the best that leads to superior performance on a continuous basis. It involves investigating practices inside and outside the industry for incorporation into a company's own operations (Thamhain, 1991). The basic philosophy of benchmarking has been summarized in Fig 2.1

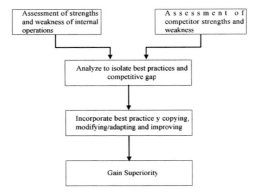

Figure 2.1 Benchmarking Objectives (Lema, 1995)

2.1.3 Types of Benchmarking

Literature does not suggest consensus on the types of benchmarking. a number of authors seem to agree on four different types. Classifications have been based mainly on approaches to benchmarking as follows (Camp 1989, Zairi 1992; Watson 1993):

♦ Internal benchmarking is performed within one organization by comparing the performance of similar business units or business processes.

♦ Competitive benchmarking is a measure of an organization's performance compared with competing organizations; studies that target specific product designs, process capabilities, or administrative methods used by a company's direct competitors; and practices or services compared with direct competitors.

♦ Functional benchmarking is an application of process benchmarking that compares particular business function in two or more organizations.

♦ Generic benchmarking is an application of functional benchmarking that compares a particular business function in two or more organizations.

15

2.1.4 Benchmarking Process Model

Benchmarking can be described as a structured process. The structure of the benchmarking process is often provided by the development of a step-by-step process model.

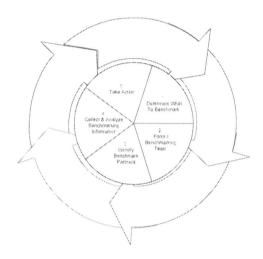

Figure 2.2 Five-stage Benchmarking Process Model
(Spendolini, 1994)

The benchmarking model is illustrated in Fig 2.2. The five process stages in benchmarking model are:

1. Determine what to benchmark.

    This first stage of the process is to identify the customers for the benchmarking information and their requirements and define the specific subjects to be benchmarked

2. Form a benchmarking team.

    Although benchmarking can be conducted by individuals, most benchmarking efforts are team activities. Specific roles and responsibilities are assigned to team members.

3. Identify benchmark partners.

    The third stage of the process involves the identification of information sources that will be used to collect benchmark information. These sources include employees of benchmarked organizations, consultants, analysts, and computerized databases. Also included in this stage is the process of identifying industry and organizational best practices.

4. Collect and analyze benchmarking information.

    Information is collected according to an established protocol and then summarized for analysis. Benchmarking information is analyzed in accordance with the original customer requirements, and recommendations for action are produced

5. Take action.

    This stage of the process is influenced by the original customer requirements and uses for the benchmarking information. The action taken may range from the production of a report or presentation to the production of a set of recommendation to the actual implementation of change based, at least in part, on the information collected in the benchmarking investigation.

    Based on the concept of benchmarking, types of benchmarking and benchmarking process model, a Benchmarking-Oriented Process Reengineering (BOPR) method can be created to facilitate project team to determine the most suitable procurement process from best-practice companies to be adapted in benchmarking company.

2.2 Semantic Similarity

2.2.1 Overview

Semantic similarity is a significant research domain in information retrieval and information integration. Several approaches to model semantic similarity compute the semantic distance between two words or two documents by definitions within ontology.

The study of semantic similarity between words has been a part of natural language processing and is a generic issue in variety applications in the areas of computational linguistics and artificial intelligence, both in the academic community and industry (Li et al., 2003). Since the first studies on interoperating information systems, progress has been made concerning syntactic (i.e., data types and formats) and structural heterogeneities (i.e., schematic integration, query languages, and interfaces) (Sheth 1992; Andrea and Egenhofer, 2003). Because of the increasing of the number of keywords and the complex relations of the contexts, the technology needed to deal successfully with these issues must focus on the semantics underlying the data used by the interoperating information systems.

Recent investigations in information retrieval and data integration have emphasized the use of ontology and semantic similarity functions as a mechanism for comparing objects that can be retrieved or integrated across heterogeneous repositories (Guarino et al., 1999; Voorhees, 1998; Smeaton and Quigley, 1996; Lee et al., 1993). Ontology is a kind of knowledge representation which describes concepts by definitions that are sufficiently detailed to capture the semantics of a specific domain. In other words, ontology can be seen as a view of the real world, and supports intentional queries regarding the content of a database. It also reflects the relevance of data by providing a declarative description of semantic information independent of the data representation (Andrea and Egenhofer, 2003).

Information system can be seen as a collection of processes that produce services; i.e., sets of related activities transforming input information into an output, to accomplish a well defined objective. Thus, more researches applied the semantic similarity to integrate the multiple, heterogeneous systems and processes relying on the data integration capability of semantic similarity analysis methods (Castano and De Antonellis 1995a, 1995b, 1997; Sciore et al. 1994).

In case of complex organizations, the number of processes to be integrated can be large, and computer-based techniques for their integration are required to perform the analysis in a systematic and semi-automatic way. Process integration can be performed following information processing view point (Galbraith, 1973). Accordingly, Castano and De Antonellos (1998) propose a framework for expressing semantic relationships between multiple information systems' processes. Via process modeling and process semantic similarity analysis manners, the processes of different information systems are successfully integrated.

Based on successful integration applications of semantic similarity, this study applied the semantic similarity analysis for benchmarking procurement process analysis between best practice companies and benchmarking company.

2.2.2 Basic Concepts for Semantic Similarity Analysis

Evaluation of process similarity is based on the terminological relationships between names specified in process descriptors. Performing a similarity-based analysis on a large number of process descriptors with an uncontrolled vocabulary is a difficult task and requires techniques for comparing names appearing in different process descriptors (Castano and De Antonellis, 1998). Major problems arising with a name-based approach are related to the fact that different processes can be characterized by the same or semantically similar entities operations, which do not necessarily have the same name. in fact, as pointed out in (Furnas, 1987) the probability of two designers picking the same name for a given entity is very low (Castano and DE Antonellis, 1998). Following recent proposals on semantic heterogeneity in multi database systems, a semantic dictionary is introduced, where names of semantic similar entities and operations are grouped under the same concept (Bright, Hurson and Pakzad, 1994; Hammer and McLeod, 1993; Castano and De Antonellis, 1998). The semantic dictionary provides capabilities to manage duplicate descriptors for processes, resulting from using different names for denoting the same entity or operation. The dictionary is organized into concept hierarchies by means of the generalization and aggregation abstraction mechanisms (Castano and De Antonellis, 1998).

The architecture of the semantic dictionary is shown in Figure 2.2. A concept hierarchy is defined in the dictionary, where the concepts at the top of the hierarchy are either more general than the lower level concepts (generalization abstraction) or are composed of them (aggregation abstraction). Each concept at the bottom level the hierarchy has associated a cluster of names corresponding to entities and operations that are semantically similar (Castano and De Antonellis, 1998). Names contained in clusters are entity and operation names used in process descriptors.

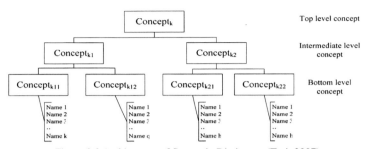

Figure 2.3 Architecture of Semantic Dictionary (Tsai, 2007)

In the semantic dictionary, concepts are connected by means of hierarchical links. In addition, a predefined link exists between a bottom-level concept and each name in the cluster associated with this concept. To operationally evaluate affinity, we assign an affinity strength (δ) to hierarchical links in the semantic dictionary. Two names in process descriptors can have affinity if they refer to a common concept in the dictionary hierarchies, that is, if a path of length l, with $l \geq 1$, denoted by the symbol "$\rightarrow^l$", exists between them in the semantic dictionary. The affinity of two names coincides with the strength of the path between them. The strength of a path "$\rightarrow^l$" is computed using the monotonic function as Equation 2.1 shows (Castano and De Antonellis, 1998).

$$NA(n_i, n_j) = \begin{cases} 1 & if\,(n_i = n_j) \\ \delta^l & if\,(n_i \neq n_j) \wedge (n_i \subseteq C_k, n_j \subseteq C_k), l \geq 1 \\ 0 & otherwise \end{cases} \dots\dots\dots\dots\dots \text{(2.1)}$$

In the Equation 2.1, $NA(n_i,n_j)$ presents the similar degree of the entity n_i and the entity n_j. δ is the affinity strength, usually set as 0.8 based on the experimentation of Castano and De Antonellis (1998); l is the semantic length between n_i and n_j in the semantic dictionary and the C_k is the concept link k in the semantic dictionary.

Based on the concept of semantic dictionary and the affinity function, more process similarity functions, such as functional similarity function and information similarity function can be derived.

2.3 Trend Model

2.3.1 Overview

The trend model was proposed by Beningson in 1971, and is used for the establishment of the organizational framework for a construction project. The model is constructed based on the scope and nature of the project. The probable organization structure of the project team is investigated to reflect the interfaces due to subcontractor relationships among the members of the project team. The niches and interfaces between the members of the project team are also analyzed in the model. The original functions of the trend model include the following: (1) clearly define the relationship between different members of the project team during the execution of the project; (2) predict the interface and mutual relationships between the members of the project team and the probable problems that will be encountered during the project execution; and (3) establish the project control system within the required time frame.

Based on the successful establishment of the organization framework for a construction project, this study applies trend model concept to calculate the communication resistance of best-practice processes occurred in the organization structure of benchmarking company.

2.3.2 Basic Concepts of Trend Model

The trend model can analyze the operation of the project team based on concepts such as mutual reliance, uncertainty, and niche positions in the operation.

1.   Mutual Reliance

Relationships between the members of the project team can be classified into the following categories: (1) independent; (2) sequential; and (3) mutual reliance.

An independent relationship indicates that the operation could be carried out through a standard procedure independently. A sequential relationship means that this particular operation could be carried out after the completion of certain preceding operations, and the degree of reliance between these operations could be classified as medium. The conflicts among the sequential operations could be minimized through proper planning and coordination in advance. As for the mutual reliance, the degree of reliance is the highest among all of the operations; thus, the job can only be completed through good coordination between the related parties. Therefore, intensive coordination should be practiced to have a unified goal and decision prior to the commencement of the work.

2.   Uncertainty

During project execution, the risk and the impact from the external environment of each work item will not be the same. Consequently, the uncertainty of the activity time is generated and the impact to the whole construction duration is classified as low, medium, or high. For example, for the construction of breakwater, the underwater leveling work of the caisson foundation and towing the caisson are heavily dependent on sea and weather conditions. Variation in the activity duration of such work will be much higher than for casting of the caisson in the dock, so the uncertainty of these works will be higher than that of other works items in the project.

3.   Niche Position

The respective working contracts and their roles in the construction sequence determine the relative positions of members in the project's organizational structure. The influence of each work item on the progress of the project could be determined through the analysis of the conflicts and the mutual relationship between the project members (Bennigson, 1971). For example, if a work item carried out by the main contractor can only be started after the completion of the subcontractor's work, due to its contractual restriction, the subcontractor should pay much more attention to finishing his/her job in order not to cause any delay to the main contractor. If the situation is reversed, the main contractor should pay less attention to that work in interest of the subcontractor.

Based on the analysis concept of trend model, an optimal project organizational structure of benchmarking company referring form best practice companies' procurement process can be studied. However, only the mutual reliance within the process and the influence of niche positions to the project are considered here, those factors that could not be clearly defined are beyond the scope of this study.

2.3.3 Trend Model's Analysis Procedure

In the process of establishing the trend model, the trend analysis procedure is adopted to define the working procedure for the model and is illustrated as follows: (1) clearly understand the content and characteristics of the work; (2) plot the project network according to the relationship between the work items; (3) estimate the uncertainty of each work item; (4) draw the diagrams of all possible organizational structures according to the clauses stipulated in the respective contracts; (5) set up mutual relationship among the members of the project team; (6) establish the trend map to facilitate the analysis of the problems associated with the communication, coordination, and mutual understanding between the project members (the conflicts between the members should also be analyzed and a solution will be brought out accordingly); (7) propose adjustments to the project organizational structure and conduct alternative analyses.

Chapter 3

BENCHMARKING ORIENTED PROCESS REENGINEERING

3.1 Basic Description of Benchmarking Oriented Process Reengineering (BOPR)

This study aims to create a Benchmarking Oriented Process Reengineering (BOPR) method to facilitate the project team in redesigning a process. By using BOPR, project team can determine the most suitable best-practice process to be adapted in benchmarking company's process.

In addressing the BOPR problem, much is found to be unsettling. For process benchmarking, most of the benchmarking companies use self-developed methods rather than established methodologies to capture and compare process data (Poulson, 1996). In regards to BPR, there are a numerous suitable reference models to assist project teams in understanding and customizing the best-practice. These reference models are formal and semiformal description of best-practice, such as business processes, data structures and handling rules. (Scheer, 1998). To overcome these problems, a systematic analysis method to determine the most suitable best-practice processes to be learned from needs to be developed. For this reason, BOPR was developed to facilitate project team in redesigning a process.

3.2 Architecture of Benchmarking Oriented Process Reengineering

The BOPR encompasses four phases analysis method which are as follows: (1) business process modeling, (2) process similarities analysis, (3) process communication index analysis, and (4) process adaptability calculation, as shown in Figure 3.1. In the business process modeling phase, a process model providing formal representation of characteristics of process from the best practice company and benchmarking company is necessarily constructed from the beginning of process reengineering. Subsequently, the semantic similarity analysis can be calculated during the process similarity analysis. Then, the degree of communication eases of best practice process to be performed in benchmarking company is analyzed during process communication analysis. Finally, an adaptability index that represents the acceptance degrees of each best practice company for benchmarking company can be summarized in the last phase.

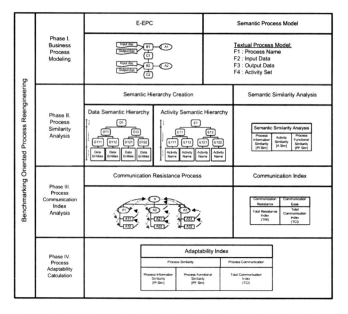

Figure 3.1 Architecture of Benchmarking Oriented Process Reengineering (BOPR)

26

Phase I. Business process modeling

A business process modeling needs to be schemed preliminarily and the relationship between the data, function and organization view of one process need to be created clearly so that a process can be considered for a given purpose, concentrating on some aspects and hiding irrelevant aspects to reduce complexity. Furthermore, a process model providing formal representation of characteristics of processes is necessarily constructed from the beginning of process reengineering. For this purpose, two tasks need to be accomplished, namely, (1) graphic process model creation, and (2) textual process model. The graphic process model is represented with eEPC diagram of ARIS modeling language (Scheer, 1998). The textual process model is mapped from graphic model into four subsets, namely, f1:{process name}, f2:{process input data}, f3{process output data} and f4{activity set} (Castano and De Antonellis, 1998). Meanwhile, each activity set is also composed of its name, input and output subsets.

Phase II. Process similarity analysis

Process similarity is one of the important factors to be considered in BOPR method. Process similarity analysis represents the corresponding relationship of data and activities between benchmarking company and best practice company. The higher the similarity of the process, the more similar the characteristics is. For the purpose, this study applies semantic similarity analysis to evaluate process information similarity (PI Sim) and process function similarity (PF Sim) between benchmarking company process and best practice company process. Therefore, a semantic hierarchy corresponding to the process models is created to depict the concept relationships of data and of activities entities, so that the semantic distance between two entities can be identified, and their semantic affinity can then be calculated. Consequently, two parameters, namely, name affinity and name set affinity, are applied to evaluate

27

process similarity.

Based on the name set affinity parameter, three types of process similarity are applied in this study, namely, Activity Similarity (*ASim*), Process Information Similarity (*PISim*), and Process Functional Similarity (*PFSim*). The process information similarity denoted by (*PISim*) is the measure of affinity degree of input and output information sets corresponding severally to two analyzed processes. Meanwhile, activity similarity denoted by (*ASim*) provides a microcosmic view form activities of a process that is necessary for advanced similarity analysis. Moreover, by summarizing all activity similarities of two processes by, the Process Functional Similarity can finally be calculated. Similar to *PISim*, the process functional similarity provides a macroscopic evaluation factor for determining two process easily to be integrated or not.

Phase III. Process Communication Index Analysis

Process communication index analysis emphasizes on organization view of a process and the main purpose is to evaluate the degree of communication ease of best-practice processes into benchmarking company organization structure so that the success of BPR implementation might be increased. By applying a trend model methodology, the degree of communication ease is evaluated. In this phase, a questionnaire survey was objectively conducted using the AHP method to obtain the resistance coefficient (Ki) of a process based on benchmarking organization structure (Min-Yuan, Cheng et al., 2003). The resistance coefficient (Ki) is used to evaluate the ease of communication for solving disputes, conflicts, or coordination problems between related parties in different layers. Based on the resistance coefficient (Ki), a Total Resistance Index (TRI) and Total Communication Index (TCI) are analyzed.

Phase IV. Process Adaptability Calculation

Process adaptability calculation summarized by process similarity analysis and process communication index analysis. The purpose of process adaptability calculation is to create an adaptability index that represents the acceptance degrees of best-practice processes for benchmarking company.

Summarily, BOPR is a systematic analysis method to determine the most suitable process from best practice companies to be implemented in benchmarking company. Proceeding to the BOPR method, understanding the underlying process differences and distinguishing best practice process as superior will be the main focus of BOPR method. Meanwhile the organization structure of the best-practice companies is another main concern for BOPR method. According to research assumption, the most suitable process is defined as a process which has similar characteristics with benchmarking company and low communication resistance when it is performed in benchmarking company. Thus, by using the BOPR method, the adaptability index (AI) which is summarized by PI Sim and PF Sim, can be calculated to determine the most suitable best-practice processes to be adapted in benchmarking company. The higher the AI is, the more suitable the process is for benchmarking company.

Chapter 4

BENCHMARKING ORIENTED PROCESS REENGINEERING
METHOD DEVELOPMENT

The purpose of this study is to determine the best-to-learn process from the collected best-practice companies for the benchmarking company. For this purpose, this study proposed the Benchmarking-Oriented Process Reengineering (BOPR) method to determine which best-practice process is most suitable to be adapted in the benchmarking company. By using the BOPR method, the Adaptability Index (AI) can be calculated to represent the acceptance degree of each best-practice process. The higher the AI is, the more suitable the process is. Accordingly, the project team can determine the best-to-learn process in accordance with the evaluated AI value.

Since this study assumes that the most suitable process has most similar characteristics with benchmarking process and can be executed smoothly when it will be performed in benchmarking company, AI considers two factors; namely (1) process similarity and (2) process communication index as Figure 4.1 shows. Process similarity is summarized by process information similarity (PI Sim) and process functional similarity (PF Sim) which expresses the similar characteristics between best-practice processes and benchmarking process; the process communication index evaluates the degree of communication ease from best-practice processes when it will be adapted in benchmarking company. To evaluate process similarity and process communication index, semantic similarities and trend model concept were applied respectively to develop BOPR method as shown in Figure 4.2.

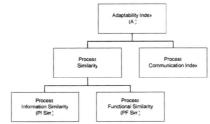

Figure 4.1 Adaptability Index (AI) Hierarchy.

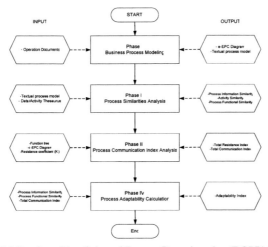

Figure 4.2 Benchmarking Oriented Process Reengineering (BOPR) Method

Figure 4.2 expresses the procedure of BOPR which includes four phases; namely

(1) business process modeling, (2) process similarities analysis, (3) process

communication analysis, and (4) process adaptability calculation. In the business

process modeling phase, a process model providing formal representation of

characteristics from the best-practice processes and benchmarking process is

necessarily constructed from the beginning of process reengineering. Subsequently,

the degree of process information similarity (PI Sim), activity similarity (A Sim), and

process functional similarity (PF Sim) were calculated in the process similarity

analysis. Then, the degree of communication eases of best practice process to be

performed in benchmarking company is analyzed during process communication analysis to calculate the total communication index (TCI) of each best practice process. Finally, an adaptability index that represents the acceptance degrees of each best practice company for benchmarking company can be summarized in the last phase.

To validate BOPR feasibility, this research further illustrated a real case study of procurement process derived from the construction company "A" which is set to be the benchmarking company and three well-performed construction companies which are set to be best-practice companies. In the case study, the BOPR method was implemented for the construction company "A" to assist the project team in determining the best-to-learn paradigm for their benchmarking project.

4.1 Business Process Modeling

In process reengineering, one of the most difficult and important tasks is to identify a company's processes. Therefore, the primary purpose of business process modeling is to develop a systematic process to assist companies in clarifying and establishing their management process (Hammer and Champy 1993). For the purpose, two tasks need to be accomplished: (1) activity data collection, which investigates an activity's functions and compiles information on the original organization, and (2) process modeling, which creates the original process model based on the collected data.

Step 1. Activity data collection

To map a business process model from function–based to process-based organization, detailed information on the activities of functional departments is required. Moreover, formal documents such as ISO descriptions and specifications or internal-auditing documents are also required to fully establish the relationships between and regulations affecting process activities.

32

Taking the case study as an example, since the case study focuses on procurement process, only the data related to procurement process need to be collected in this stage. The relevant data and activity information regarding procurement process from the construction company "A" and three best practice companies were collected to create procurement process model for each company.

Step 2. Process modeling

Two process models, namely (1) e-EPC Diagram, and (2) textual process model, are necessary to be created in this step. Based on the collected data, a process model can be represented by using ARIS modeling language eEPC diagram and textual process model. ARIS modeling tool is used to create the original process model based on the collected data; the textual process model is used to create data-oriented process model which is necessary for the process similarity analysis in the next phase.

In the case study, detailed procurement process models from three best-practice companies and the construction company "A" are created. Figure 4.3, Figure 4.4, Figure 4.5 and Figure 4.6 show the e-EPC diagram of the procurement process models from the benchmarking company and the three best-practice companies.

After e-EPC diagram was created, a textual process model which focuses on representing process data information needs to be created subsequently to be the input of the similarity analysis. The textual process model is simplified from the graphic model e-EPC in order to summarize naming the operations and present data entities within the process. Therefore, four subsets are included; namely, $f1$:{process name}, $f2$:{process input data set}, $f3${process output data set} and $f4${activity set}. Each activity set is composed of activity names and input and output subsets. Figure 4.7, Figure 4.8, Figure 4.9 and Figure 4.10 show the textual process model from the benchmarking company and three best-practice companies derived from the e-EPC diagram of Figure 4.3 – 4.6.

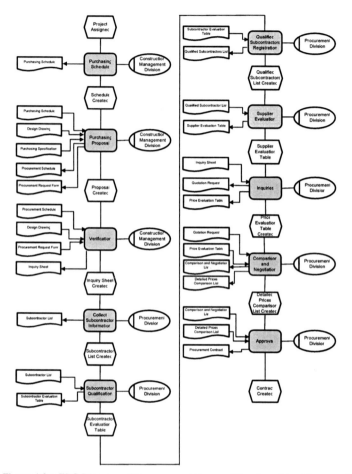

Figure 4.3 e-EPC Diagram of Procurement Process [Benchmarking Company]

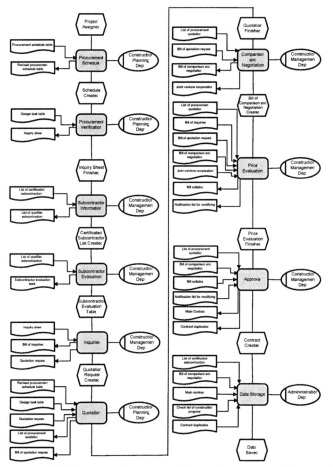

Figure 4.4 e-EPC Diagram of Procurement Process [Best practice Company 1]

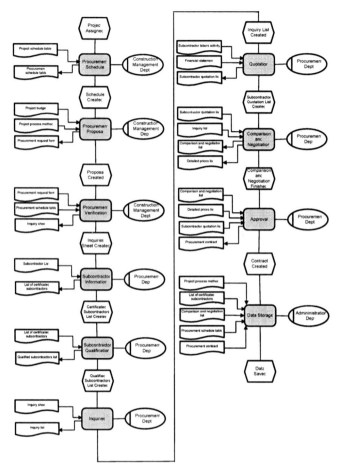

Figure 4.5 e-EPC Diagram of Procurement Process [Best practice Company 2]

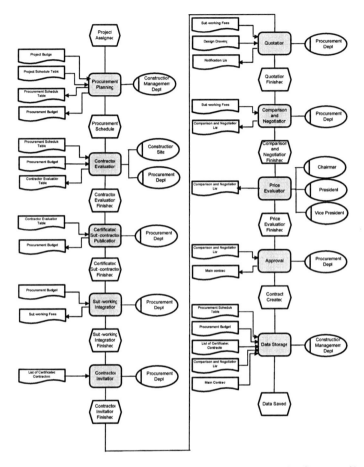

Figure 4.6 e-EPC Diagram of Procurement Process [Best practice Company 3]

Textual Process Model of Procurement Process:

F1   : Procurement Process

F2   :{Purchasing   schedule,   Design   drawing,   Purchasing   specification, Subcontractor
        list}

F3   :{Procurement   schedule,   Procurement   request   form,   Inquiry   sheet, Subcontractor
        evaluation table, Qualified subcontractor list, Supplier evaluation table,
        Quotation request, Price evaluation table. Comparison and negotiation list,
        Detailed prices comparison list, Procurement contract}

F4   : <Purchasing Schedule, {Purchasing schedule}>
        <Purchasing Proposal, {Purchasing schedule, Design drawing, Purchasing
         specification}, {Procurement schedule, Procurement request form}>
        <Verification, {Procurement schedule, Design drawing, Procurement request
         form}{Inquiry sheet}>
        <Collect Subcontractor Information, {Subcontractor list}>
        <Subcontractor Qualification, {Subcontractor list}, {Subcontractor evaluation
         table}>
        <Qualified Subcontractor Registration, {Subcontractor evaluation table},
         {Qualified subcontractor list}>
        <Supplier Evaluation, {Qualified subcontractor list}, {Supplier evaluation
         table}>
        <Inquiries, {Inquiry sheet}, {Quotation request, Price evaluation table}>
        <Comparison and Negotiation, {Quotation request, Price evaluation table},
         {Comparison and negotiation list, Detailed prices comparison list}>
        <Approval, {Comparison and negotiation list, Detailed prices comparison list},
         {Procurement contract}

Figure 4.7 Textual Process Model of Procurement Process [Benchmarking Company]
(derived from Figure 4.3).

Textual Process Model of Procurement Process:

F1   : Procurement Process

F2   : {Procurement schedule table, Design task table, List of certificated contractor, Check list of construction progress}

F3   : {Revised procurement schedule table, Inquiries sheet, List of qualified Subcontractors, Subcontractor evaluation table, Bill of inquiries, Quotation request, List of procurement quotation, Bill of quotation request, Bill of comparison and negotiation, Joint-venture cooperation, Bill collation, Notification list for modifying, Main contract, Contract duplication}

F4   :   <Procurement   Schedule,   {Procurement   schedule   table},   {Revised procurement
        schedule table}>
        <Procurement Verification, {Design task table}, {Inquiry sheet}>
        <Subcontractor Information, {List of certificated subcontractors}, {List of qualified subcontractors}>
        <Subcontractor Evaluation, {List of qualified subcontractors}, {Subcontractor evaluation table}>
        <Inquiries, {Inquiries sheet}, {Bill of inquiries, Quotation request}>
        <Quotation, {Revised procurement schedule table, Design task table, Quotation request}, {List of procurement quotation, Bill of quotation request}>
        <Comparison and Negotiation, {List of procurement quotation, Bill of quotation request}, {Bill of comparison and negotiation, Joint venture cooperation}>
        <Price Evaluation, {List of procurement quotation, Bill of inquiries, Bill of quotation request, Bill of comparison and negotiation, Joint venture cooperation}, {Bill collation, Notification list for modifying}>
        <Approval, {List of procurement quotation, Bill of comparison and negotiation, Bill Collation, Notification list for modifying}, {Main contract, Contract duplication}>
        <Data Storage, {List of certificated subcontractors, Bill of comparison and negotiation, Main contract, Check list of construction progress, Contract duplication}>

Figure 4.8 Textual Process Model of Procurement Process [Best practice Company 1] (derived from Figure 4.4).

Textual Process Model of Procurement Process:	
F1	: Procurement Process
F2	: {Project schedule table, Project budget, Project process method, Subcontractor list, Subcontractor labors activity, Financial statement}
F3	: {Procurement schedule table, Procurement request form, Inquiry sheet, List of Certificated subcontractors, Qualified subcontractors list, Inquiry list, Subcontractor Quotation list, Comparison and negotiation list, Detailed prices list, Procurement contract}
F4	: <Procurement Schedule, {Project schedule table}, {Procurement schedule table}>
	<Procurement Proposal, {Project budget, Project process method}, {Procurement request form}>
	<Procurement Verification, {Procurement request form, Procurement schedule table}, {Inquiry sheet}>
	<Subcontractor Information, {Subcontractor list}, {List of certificated subcontractors}>
	<Subcontractor Qualification, {List of certificated subcontractors}, {Qualified subcontractors list}>
	<Inquiries, {Inquiry sheet}, {Inquiry list}>
	<Quotation, {Subcontractor labors activity, Financial statement}, {Subcontractor quotation list}>
	<Comparison and Negotiation, {Subcontractor quotation list, Inquiries list}, {Comparison and negotiation list, Detailed prices list}>
	<Approval, {Comparison and negotiation list, Detailed process list, Subcontractor quotation list}, {Procurement contract}>
	<Data Storage, {Project process method, List of certificated subcontractors, Comparison and negotiation list, Procurement schedule table, Procurement contract}>

Figure 4.9 Textual Process Model of Procurement Process [Best practice Company 2]
(derived from Figure 4.5).

```
Textual Process Model of Procurement Process:
F1    : Procurement Process
F2    : {Project budget, Project schedule table, Design drawing}
F3    : {Procurement schedule table, Procurement budget, Contractor evaluation
table,
        List of certificated contractor, Sub-working fees, Notification list, Comparison
        And negotiation list, Main contract}
F4    : <Procurement Planning, {Project budget, Project schedule table},
        {Procurement schedule table, Procurement budget}>
      <Contractor Evaluation, {Procurement schedule table, Procurement budget},
      {Contractor evaluation table}>
      <Certificated Sub-contractor Publication, {Contractor evaluation table}, {List of
      certificated sub-contractors}>
       <Sub-working Integration, {Procurement budget}, {Sub-working fees}>
      <Contractor Invitation, {List of certificated sub-contractor}>
      <Quotation, {Sub-working fees, Design drawing}, {Notification list}>
      <Comparison and Negotiation, {Sub-working fees}, {Comparison and
      Negotiation list}>
      <Price Evaluation, {Comparison and negotiation list}>
      <Approval, {Comparison and negotiation list}, {Main contract}>
      <Data Storage, {Procurement budget, Procurement schedule budget, List of
      certificated sub-contractors, Comparison and negotiation list, Main contract}>
```

Figure 4.10 Textual Process Model of Procurement Process [Best practice Company 3]
(derived from Figure 4.6)

Based on the textual process model, the semantic hierarchy for similarity analysis in the next phase can be generated.

4.2 Process Similarity Analysis

Process similarity is one of the important factors to be considered in BOPR method. The primary purpose of process similarity analysis is to evaluate process information similarity (PI Sim) and process functional similarity (PF Sim) between the benchmarking company and the best-practice companies. The higher the PI Sim is, the more similar input or output data the two processes have; similarly, the higher the PF Sim is, the more similar activities the two processes have. Summarizing the PI Sim and PF Sim, the process similarity can be finally calculated to express the commonalities between two processes. For the purpose, this study applies semantic similarity analysis to evaluate PI Sim and PF Sim between benchmarking company and best-practice company processes, because of ontological evaluation ability of the semantic similarity analysis.

Based on the semantic similarity analysis philosophy, two primary tasks are involved in the process similarity analysis. First, the semantic hierarchy needs to be created to provide the semantic distance between each two analyzed entities within the process. Then, based on the semantic hierarchy creation, the similarity analysis can be calculated.

4.2.1 Semantic Hierarchy Creation

Semantic hierarchy is the foundation upon which semantic similarity analysis is employed to identify similarities between processes in terms of data and functions. Semantic hierarchy depth represents the semantic distance between two analyzed entities. Degree of semantic similarity can be calculated by referencing semantic distance.

Semantic hierarchy, when clustered with similar data entities and activity functions, provides semantic distances for semantic similarity analysis. Two semantic hierarchies for, respectively, data and activity were created. The data semantic hierarchy gathers data entities from the represented process models created in the first phase of BOPR. The activity semantic hierarchy groups similar activity names within process models into several clusters. The semantic hierarchy clusters similarities into two groups. The length of the path between two clusters (semantic distance) is identified according to hierarchical structure and the similarity of two entities is calculated. The steps followed in creating data and activity semantic hierarchy are described below.

4.2.1.1   Data Semantic Hierarchy Creation

Data semantic hierarchy represents the semantic distance between two analyzed data entities. By referencing the semantic distance, the degree of data semantic similarity can be calculated. Three steps are involved in the creation of a data semantic hierarchy.

Step 1. Create process model data thesaurus

To create a semantic hierarchy, this study used a data thesaurus to gather all data entities within selected processes and represent the semantic relationships between data entities. Figure 4.11, Figure 4.12, Figure 4.13 shows an example of a data thesaurus based on the case study procurement process. The symbols n_i and n_j represent data entity names, and l_{ij} represents the linkage between n_i and n_j. A linkage is a triplet $l_{ij}=(r_{ij}, a, k)$, where r_{ij} represents semantic relationship type; a represents the affinity associated with the relationship, and k represents the number of occurrences of the relationship type for the considered name pair (a large k provides evidence for a relationship between a pair of data entities). Following Castano and Antonellis (1998), this study applies three types of semantic relationships, namely SYN, BT/NT, and RT. The definition of each is described below:

- SYN (SYNonymy): two entities are synonymies when they can replace each other in all processes without changes in meaning. The relationship affinity a_{SYN} = 1.

- BT/NT (Broader/Narrow Terms): two names are defined as a BT (or NT) relationship when one has a more (or less) the same general meaning as the other. The relationship affinity $a_{BT} = a_{NT} = 0.8$.

43

- *RT* (Related Terms): or positive association, defined between names that are generally used in the same context. The relationship affinity $a_{RT} = 0.5$.

These three relationships are defined as explicit relationships. Denoted by the symbol "$\Re$", because they can be identified subjectively. Moreover, when two processes are analyzed, some affinities may be found between entities within two different processes that result from their participation in a third entity. This kind of relationship, denoted by the symbol "$\Rightarrow$", is defined as an implicit relationship. Using the data thesaurus shown in Figure 4.10 as an example, the entity, A04 Subcontractor list in the 4th column of benchmarking company procurement process, incorporates more information than does the B03 List of certificated contractor entity in the 3rd row of the best practice company 1 procurement process. Their relationship is thus defined as "BT". The next step is to identify other entities in the benchmarking company procurement process that share a relationship with the best practice company 1 procurement process. For example, the Subcontractor list (A04) shares a Related Term "RT" relationship with the Inquiry sheet (A07). The intersection of "A07" and "B03" is defined as an implicit (BT*RT) relationship. Relationships in the rest of the table were determined in the same manner.

Columns: Nj (Benchmarking Company)

Ni	ly	A01 Purchasing Schedule	A02 Design Drawing	A03 Purchasing Specification	A04 Subcontractor List	A05 Procurement Schedule	A06 Procurement Request Form	A07 Inquiry Sheet	A08 Subcontractor Evaluation Table	A09 Qualified Subcontractor List	A10 Supplier Evaluation Table	A11 Quotation Request	A12 Price Evaluation Table	A13 Comparison and Negotiation List	A14 Detailed Prices Comparison List	A15 Procurement Contract
Benchmarking Company A01	Purchasing Schedule	(syn,1,1)	-	(rt,0.5,1)	-	(bt,0.8,1)	-	-	-	-	-	-	-	-	-	(rt,0.5,1)
A02	Design Drawing	-	(syn,1,1)	(rt,0.5,1)	-	(rt,0.5,1)	-	-	-	-	-	-	-	-	-	-
A03	Purchasing Specification	(rt,0.5,1)	(rt,0.5,1)	(syn,1,1)	-	(nt,0.8,1)	-	-	-	(rt,0.5,1)	-	-	-	-	-	-
A04	Subcontractor List	-	-	-	(syn,1,1)	-	(nt,0.8,1)	(rt,0.5,1)	(bt,0.8,1)	(rt,0.5,1)	(rt,0.5,1)	-	-	-	-	-
A05	Procurement Schedule	(bt,0.8,1)	(rt,0.5,1)	(nt,0.8,1)	-	(syn,1,1)	(rt,0.5,1)	(rt,0.5,1)	-	-	-	-	(rt,0.5,1)	-	-	(rt,0.5,1)
A06	Procurement Request Form	-	-	-	(nt,0.8,1)	(rt,0.5,1)	(syn,1,1)	(rt,0.5,1)	-	-	-	(bt,0.8,1)	-	(bt,0.8,1)	-	-
A07	Inquiry Sheet	-	-	-	(rt,0.5,1)	(rt,0.5,1)	(rt,0.5,1)	(syn,1,1)	-	-	-	(nt,0.8,1)	-	(nt,0.8,1)	-	-
A08	Subcontractor Evaluation Table	-	-	-	(bt,0.8,1)	-	-	-	(syn,1,1)	(bt,0.8,1)	(rt,0.5,1)	-	-	(rt,0.5,1)	-	-
A09	Qualified Subcontractor List	-	-	(rt,0.5,1)	(rt,0.5,1)	-	-	-	(bt,0.8,1)	(syn,1,1)	(rt,0.5,1)	-	-	(rt,0.5,1)	-	-
A10	Supplier Evaluation Table	-	-	-	(rt,0.5,1)	-	-	-	(rt,0.5,1)	(rt,0.5,1)	(syn,1,1)	(rt,0.5,1)	-	(rt,0.5,1)	-	-
A11	Quotation Request	-	-	-	-	-	(bt,0.8,1)	(nt,0.8,1)	-	-	(rt,0.5,1)	(syn,1,1)	(rt,0.5,1)	(rt,0.5,1)	(rt,0.5,1)	(rt,0.5,1)
A12	Price Evaluation Table	-	-	-	-	(rt,0.5,1)	-	-	-	-	-	(rt,0.5,1)	(syn,1,1)	(nt,0.8,1)	(nt,0.8,1)	(rt,0.5,1)
A13	Comparison and Negotiation List	-	-	-	-	-	(bt,0.8,1)	(nt,0.8,1)	(rt,0.5,1)	(rt,0.5,1)	(rt,0.5,1)	(rt,0.5,1)	(nt,0.8,1)	(syn,1,1)	(rt,0.5,1)	(rt,0.5,1)
A14	Detailed Prices Comparison List	-	-	-	-	-	-	-	-	-	-	(rt,0.5,1)	(nt,0.8,1)	(rt,0.5,1)	(syn,1,1)	(rt,0.5,1)
A15	Procurement Contract	(rt,0.5,1)	-	-	-	(rt,0.5,1)	-	-	-	-	-	(rt,0.5,1)	(rt,0.5,1)	(rt,0.5,1)	(rt,0.5,1)	(syn,1,1)
Best-practice Company (1) B01	Procurement Schedule Table	(rt,0.5,1)	implicit (syn*rt)	implicit (syn*rt) (rt*rt)	-	(syn,1,1)	implicit (syn*rt)	implicit (syn*rt)	-	-	-	-	implicit (syn*rt)	-	-	implicit (rt*rt) (syn*rt) (syn*rt)
B02	Design Task Table	implicit (rt*rt)	(syn,1,1)	implicit (syn*rt)	-	implicit (bt*rt) (syn*rt)	-	implicit (syn*rt)	-	-	-	implicit (nt*rt)	-	-	-	-
B03	List of Certificated Contractor	-	-	-	(bt,0.8,1)	-	implicit (bt*nt)	implicit (bt*rt)	implicit (syn*nt) (bt*bt)	implicit (syn*bt) (bt*rt)	implicit (nt*nt) (bt*rt)	implicit (syn*rt)	-	-	-	-
B04	Check list of Construction Progress	implicit (rt*rt)	-	-	-	implicit (bt*rt)	-	-	implicit (syn*rt)	-	implicit (nt*rt)	-	-	implicit (rt*rt)	implicit (nt*rt)	implicit (syn*rt)
B05	Revised Procurement Schedule Table	(rt,0.5,1)	implicit (syn*rt) (bt*rt)	implicit (rt*rt) (bt*nt)	-	(bt,0.8,1)	implicit (bt*rt)	implicit (bt*rt)	-	-	-	implicit (syn*rt)	implicit (bt*rt)	-	-	implicit (rt*rt) (bt*rt) (syn*rt)
B06	Inquiry Sheet	-	implicit (syn*rt)	-	implicit (syn*rt)	implicit (syn*rt)	implicit (syn*rt)	(syn,1,1)	-	-	-	implicit (nt*rt) (syn*nt) (syn*nt)	-	implicit (syn*nt)	-	-
B07	List of Qualified Subcontractors	-	-	implicit (syn*rt)	(nt,0.8,1)	-	implicit (nt*nt)	implicit (nt*rt)	implicit (bt*nt) (syn*nt) (syn*bt)	(syn,1,1)	implicit (bt*nt) (nt*rt) (syn*rt)	implicit (syn*rt) (nt*rt)	-	implicit (syn*rt)	-	implicit (syn*rt)
B08	Subcontractor Evaluation Table	-	-	-	implicit (bt*nt) (nt*rt) (nt*rt) (syn*bt)	-	-	-	(syn,1,1)	implicit (nt*rt) (syn*nt) (syn*bt)	(nt,0.8,1)	implicit (nt*rt)	implicit (rt*rt)	implicit (nt*rt) (syn*rt) (syn*rt)	-	implicit (syn*rt)
B09	Bill of Inquiries	-	-	-	implicit (nt*rt)	implicit (nt*rt)	implicit (nt*rt)	(nt,0.8,1)	-	-	-	implicit (bt*nt) (nt*nt) (syn*bt)	-	implicit (rt*rt) (nt*nt)	implicit (rt*rt)	-
B10	Quotation Request	implicit (rt*rt)	-	-	implicit (bt*rt) (nt*rt)	implicit (bt*rt)	implicit (syn*bt)	implicit (bt*nt) (syn*nt) (syn*nt)	-	implicit (syn*rt)	implicit (syn*rt)	(syn,1,1)	implicit (syn*rt)	implicit (rt*rt) (syn*rt)	implicit (syn*rt)	implicit (syn*rt)
B11	List of Procurement Quotation	-	implicit (syn*rt)	-	-	-	-	implicit (syn*rt) (nt*rt)	-	-	-	implicit (syn*rt) (nt*rt)	implicit (rt*rt^1/2)	implicit (rt*rt^1/2) (rt*rt)	implicit (rt*rt)	implicit (syn*rt)
B12	Bill of Quotation Request	-	implicit (syn*rt)	-	implicit (nt*rt)	-	implicit (bt*nt)	implicit (bt*nt) (nt*nt) (syn*rt)	-	implicit (syn*rt)	implicit (nt*rt)	(nt,0.8,1)	implicit (rt*rt) (nt*rt)	implicit (nt*rt) (rt*rt) (syn*rt)	implicit (rt*rt) (nt*rt)	implicit (nt*rt) (syn*rt)
B13	Bill of Comparison and Negotiation	-	-	-	-	implicit (rt*rt)	implicit (syn*bt)	implicit (syn*nt)	implicit (syn*rt) (syn*rt)	implicit (syn*rt)	implicit (syn*rt) (nt*rt)	implicit (rt*rt) (rt*rt) (syn*rt)	(rt,0.5,1)	(syn,1,1)	implicit (rt*rt) (nt*rt) (syn*rt)	implicit (rt*rt) (syn*rt) (syn*rt)
B14	Joint Venture Cooperation	-	-	-	implicit (nt*rt)	-	-	-	implicit (syn*rt)	implicit (syn*rt)	implicit (nt*rt)	implicit (nt*rt)	implicit (rt*rt)	implicit (rt*rt) (syn*rt)	-	implicit (syn*rt)
B15	Bill Collation	-	-	-	-	-	implicit (bt*rt)	implicit (nt*rt) (nt*rt)	implicit (rt*rt)	implicit (rt*rt)	implicit (rt*rt)	implicit (nt*rt) (rt*rt) (syn*rt)	implicit (rt*rt) (nt*rt)	(rt,0.5,1)	implicit (rt*rt) (rt*rt)	implicit (rt*rt) (syn*rt)
B16	Notification List for Modifying	-	-	-	-	-	-	implicit (nt*rt)	-	-	-	implicit (rt*rt) (rt*rt) (rt*rt)	implicit (rt*rt) (nt*rt)	implicit (rt*rt) (rt*rt) (syn*rt)	(rt,0.5,1)	implicit (rt*rt) (syn*rt)
B17	Main Contract	implicit (rt*nt) (rt*rt) (syn*rt)	-	-	implicit (nt*rt)	implicit (bt*rt) (syn*rt) (syn*rt)	-	-	implicit (syn*rt)	implicit (syn*rt)	implicit (nt*rt)	implicit (syn*rt) (nt*rt)	implicit (rt*rt) (syn*rt)	implicit (rt*rt) (syn*rt) (syn*rt)	implicit (rt*rt) (syn*rt)	(syn,1,1)
B18	Contract Duplication	implicit (rt*rt) (rt*rt)	-	-	implicit (nt*rt)	implicit (syn*rt) (bt*rt)	-	-	implicit (syn*nt)	implicit (syn*nt)	implicit (nt*rt)	implicit (rt*rt)	implicit (rt*rt)	implicit (rt*rt) (syn*rt)	implicit (rt*rt)	implicit (syn*bt)

Columns: Nj (Best-practice Company (1))

Ni	ly	B01 Procurement Schedule Table	B02 Design Task Table	B03 List of Certificated Contractor	B04 Check List of Construction Progress	B05 Revised Procurement Schedule Table	B06 Inquiry Sheet	B07 List of Qualified Subcontractors	B08 Subcontractor Evaluation Table	B09 Bill of Inquiries	B10 Quotation Request	B11 List of Procurement Quotation	B12 Bill of Quotation Request	B13 Bill of Comparison and Negotiation	B14 Joint Venture Cooperation	B15 Bill Collation	B16 Notification List for Modifying	B17 Main Contract	B18 Contract Duplication
Benchmarking Company A01	Purchasing Schedule	(rt,0.5,1)	implicit (rt*rt)	-	implicit (rt*rt)	(rt,0.5,1)	-	-	-	-	implicit (rt*rt)	-	-	-	-	-	-	implicit (rt*nt) (rt*rt) (syn*rt)	implicit (rt*rt) (rt*rt)
A02	Design Drawing	implicit (syn*rt)	(syn,1,1)	-	-	implicit (syn*rt) (bt*rt)	implicit (syn*rt)	-	-	-	-	implicit (syn*rt)	implicit (syn*rt)	-	-	-	-	-	-
A03	Purchasing Specification	implicit (syn*nt) (rt*rt)	implicit (syn*rt)	-	-	implicit (rt*rt) (bt*nt)	-	implicit (syn*rt)	-	-	-	-	-	-	-	-	-	-	-
A04	Subcontractor List	-	-	(bt,0.8,1)	-	-	implicit (syn*rt)	(nt,0.8,1)	implicit (bt*nt) (nt*rt) (nt*rt) (syn*bt)	implicit (nt*rt)	implicit (bt*rt) (nt*rt)	-	implicit (nt*rt)	-	implicit (nt*rt)	-	-	implicit (nt*rt)	implicit (nt*rt)
A05	Procurement Schedule	(syn,1,1)	implicit (bt*rt) (syn*rt)	-	implicit (bt*rt)	(bt,0.8,1)	implicit (syn*rt)	-	-	implicit (nt*rt)	implicit (bt*rt)	-	-	implicit (rt*rt)	-	-	-	implicit (bt*rt) (syn*rt) (syn*rt)	implicit (syn*rt) (bt*rt)
A06	Procurement Request Form	implicit (syn*rt)	-	implicit (bt*nt)	-	implicit (bt*rt)	implicit (syn*rt)	implicit (nt*nt)	-	implicit (nt*rt)	implicit (syn*bt)	-	implicit (bt*nt)	implicit (syn*bt)	-	implicit (bt*rt)	-	-	-
A07	Inquiry Sheet	implicit (syn*rt)	implicit (syn*rt)	implicit (bt*rt)	-	implicit (bt*rt)	(syn,1,1)	implicit (nt*rt)	-	(nt,0.8,1)	implicit (bt*nt) (syn*nt) (syn*nt)	implicit (syn*rt) (nt*rt)	implicit (bt*nt) (nt*nt) (syn*rt)	implicit (syn*nt)	-	implicit (nt*rt) (nt*rt)	implicit (nt*rt)	-	-
A08	Subcontractor Evaluation Table	-	-	implicit (syn*nt) (bt*bt)	implicit (syn*rt)	-	-	implicit (bt*nt) (syn*nt) (syn*bt)	(syn,1,1)	-	-	-	-	implicit (syn*rt) (syn*rt)	implicit (syn*rt)	implicit (rt*rt)	-	implicit (syn*rt)	implicit (syn*rt)
A09	Qualified Subcontractor List	-	-	implicit (syn*bt) (bt*rt)	-	-	-	(syn,1,1)	implicit (nt*rt) (syn*nt) (syn*bt)	-	implicit (syn*rt)	-	implicit (syn*rt)	implicit (syn*rt)	implicit (syn*rt)	implicit (rt*rt)	-	implicit (syn*rt)	implicit (syn*rt)
A10	Supplier Evaluation Table	-	-	implicit (nt*nt) (bt*rt)	implicit (nt*rt)	-	-	implicit (nt*nt) (nt*rt) (syn*rt)	(nt,0.8,1)	-	implicit (syn*rt)	-	implicit (nt*rt)	implicit (syn*rt) (nt*rt)	implicit (nt*rt)	implicit (rt*rt)	-	implicit (nt*rt)	implicit (nt*rt)
A11	Quotation Request	-	implicit (nt*rt)	implicit (syn*rt)	-	implicit (syn*rt)	implicit (nt*rt) (syn*nt) (syn*nt)	implicit (syn*rt) (nt*rt)	implicit (nt*rt)	implicit (bt*nt) (nt*nt) (syn*bt)	(syn,1,1)	implicit (syn*rt) (nt*rt)	(nt,0.8,1)	implicit (rt*rt) (rt*rt) (syn*rt)	implicit (nt*rt)	implicit (nt*rt) (rt*rt) (syn*rt)	implicit (rt*rt) (nt*rt)	implicit (syn*rt) (nt*rt)	implicit (nt*rt)
A12	Price Evaluation Table	implicit (syn*rt)	-	-	-	implicit (bt*rt)	-	-	implicit (rt*rt)	-	implicit (syn*rt)	implicit (rt*rt^1/2)	implicit (rt*rt) (nt*rt)	(rt,0.5,1)	implicit (rt*rt)	implicit (rt*rt) (nt*rt)	implicit (rt*rt) (nt*rt)	implicit (rt*rt) (syn*rt)	implicit (rt*rt)
A13	Comparison and Negotiation List	-	-	-	implicit (rt*rt)	-	implicit (syn*nt)	implicit (syn*rt)	implicit (nt*rt) (syn*rt) (syn*rt)	implicit (rt*rt) (nt*nt)	implicit (rt*rt) (syn*rt)	implicit (rt*rt^1/2) (rt*rt)	implicit (nt*rt) (rt*rt) (syn*rt)	(syn,1,1)	implicit (rt*rt) (syn*rt)	(rt,0.5,1)	implicit (rt*rt) (rt*rt) (syn*rt)	implicit (rt*rt) (syn*rt) (syn*rt)	implicit (rt*rt) (syn*rt)
A14	Detailed Prices Comparison List	-	-	-	implicit (rt*rt)	-	-	-	-	implicit (rt*rt)	implicit (syn*rt)	implicit (rt*rt)	implicit (rt*rt) (nt*rt)	implicit (rt*rt) (nt*rt) (syn*rt)	-	implicit (rt*rt) (rt*rt)	(rt,0.5,1)	implicit (rt*rt) (syn*rt)	implicit (rt*rt)
A15	Procurement Contract	implicit (rt*rt) (syn*rt) (syn*rt)	-	-	implicit (syn*rt)	implicit (rt*rt) (bt*rt) (syn*rt)	-	implicit (syn*rt)	implicit (syn*rt)	-	implicit (syn*rt)	implicit (syn*rt)	implicit (nt*rt) (syn*rt)	implicit (rt*rt) (syn*rt) (syn*rt)	implicit (syn*rt)	implicit (rt*rt) (syn*rt)	implicit (rt*rt) (syn*rt)	(syn,1,1)	implicit (syn*bt)
Best-practice Company (1) B01	Procurement Schedule Table	(syn,1,1)	-	-	-	(bt,0.8,1)	-	-	-	-	-	-	-	-	-	-	-	(rt,0.5,1)	(rt,0.5,1)
B02	Design Task Table	-	(syn,1,1)	-	-	(rt,0.5,1)	(rt,0.5,1)	-	-	-	-	(rt,0.5,1)	(rt,0.5,1)	-	-	-	-	-	-
B03	List of Certificated Contractor	-	-	(syn,1,1)	-	-	-	(bt,0.8,1)	(nt,0.8,1)	-	(rt,0.5,1)	-	-	-	-	-	-	-	-
B04	Check list of Construction Progress	-	-	-	(syn,1,1)	(rt,0.5,1)	-	-	(rt,0.5,1)	-	-	-	-	-	(rt,0.5,1)	(rt,0.5,1)	(rt,0.5,1)	(rt,0.5,1)	(rt,0.5,1)
B05	Revised Procurement Schedule Table	(bt,0.8,1)	(rt,0.5,1)	-	(rt,0.5,1)	(syn,1,1)	-	-	-	-	(rt,0.5,1)	-	-	-	-	-	-	(rt,0.5,1)	(rt,0.5,1)
B06	Inquiry Sheet	-	(rt,0.5,1)	-	-	-	(syn,1,1)	-	-	(nt,0.8,1)	(nt,0.8,1)	(rt,0.5,1)	(rt,0.5,1)	-	-	-	-	-	-
B07	List of Qualified Subcontractors	-	-	(bt,0.8,1)	-	-	-	(syn,1,1)	(nt,0.8,1)	-	(rt,0.5,1)	-	(rt,0.5,1)	-	(rt,0.5,1)	-	-	(rt,0.5,1)	(rt,0.5,1)
B08	Subcontractor Evaluation Table	-	-	(nt,0.8,1)	(rt,0.5,1)	-	-	(nt,0.8,1)	(syn,1,1)	-	-	-	-	(rt,0.5,1)	(rt,0.5,1)	-	-	(rt,0.5,1)	(rt,0.5,1)
B09	Bill of Inquiries	-	-	-	-	-	(nt,0.8,1)	-	-	(syn,1,1)	(bt,0.8,1)	(rt,0.5,1)	(bt,0.8,1)	-	-	(rt,0.5,1)	(rt,0.5,1)	-	-
B10	Quotation Request	-	-	(rt,0.5,1)	-	(rt,0.5,2)	(nt,0.8,1)	(rt,0.5,1)	-	(bt,0.8,1)	(syn,1,1)	(rt,0.5,1)	(nt,0.8,1)	-	-	(rt,0.5,1)	-	-	-
B11	List of Procurement Quotation	-	(rt,0.5,1)	-	-	-	(rt,0.5,1)	-	-	(rt,0.5,1)	(rt,0.5,1)	(syn,1,1)	(rt,0.5,1)	(rt,0.5,2)	(rt,0.5,1)	(rt,0.5,1)	(rt,0.5,1)	(rt,0.5,1)	(rt,0.5,1)
B12	Bill of Quotation Request	-	(rt,0.5,1)	-	-	-	(rt,0.5,1)	(rt,0.5,1)	-	(bt,0.8,1)	(nt,0.8,1)	(rt,0.5,1)	(syn,1,1)	(rt,0.5,1)	(rt,0.5,1)	(rt,0.5,1)	(rt,0.5,1)	(rt,0.5,1)	(rt,0.5,1)
B13	Bill of Comparison and Negotiation	-	-	-	-	-	-	-	(rt,0.5,1)	-	-	(rt,0.5,2)	(rt,0.5,1)	(syn,1,1)	(rt,0.5,1)	(rt,0.5,1)	(rt,0.5,1)	(rt,0.5,1)	(rt,0.5,1)
B14	Joint Venture Cooperation	-	-	-	(rt,0.5,1)	-	-	(rt,0.5,1)	(rt,0.5,1)	-	-	(rt,0.5,1)	(rt,0.5,1)	(rt,0.5,1)	(syn,1,1)	(rt,0.5,1)	-	(rt,0.5,1)	(rt,0.5,1)
B15	Bill Collation	-	-	-	(rt,0.5,1)	-	-	-	-	(rt,0.5,1)	(rt,0.5,1)	(rt,0.5,1)	(rt,0.5,1)	(rt,0.5,1)	(rt,0.5,1)	(syn,1,1)	(rt,0.5,1)	(rt,0.5,1)	(rt,0.5,1)
B16	Notification List for Modifying	-	-	-	(rt,0.5,1)	-	-	-	-	(rt,0.5,1)	-	(rt,0.5,1)	(rt,0.5,1)	(rt,0.5,1)	-	(rt,0.5,1)	(syn,1,1)	(rt,0.5,1)	(rt,0.5,1)
B17	Main Contract	(rt,0.5,1)	-	-	(rt,0.5,1)	(rt,0.5,1)	-	(rt,0.5,1)	(rt,0.5,1)	-	-	(rt,0.5,1)	(rt,0.5,1)	(rt,0.5,1)	(rt,0.5,1)	(rt,0.5,1)	(rt,0.5,1)	(syn,1,1)	(bt,0.8,1)
B18	Contract Duplication	(rt,0.5,1)	-	-	(rt,0.5,1)	(rt,0.5,1)	-	(rt,0.5,1)	(rt,0.5,1)	-	-	(rt,0.5,1)	(rt,0.5,1)	(rt,0.5,1)	(rt,0.5,1)	(rt,0.5,1)	(rt,0.5,1)	(bt,0.8,1)	(syn,1,1)

Figure 4.11 Data Thesaurus [Best practice company 1]

Figure 4.12 Data Thesaurus [Best practice company 2]

Benchmarking Company / Best-practice Company (3) — Data Thesaurus

Ni \ Nj		A01 Purchasing Schedule	A02 Design Drawing	A03 Purchasing Specifications	A04 Subcontractor List	A05 Procurement Schedule	A06 Procurement Request Form	A07 Inquiry Sheet	A08 Subcontractor Evaluation Table	A09 Qualified Subcontractor List	A10 Supplier Evaluation Table	A11 Quotation Request	A12 Price Evaluation Table	A13 Comparison and Negotiation List	A14 Detailed Prices Comparison List	A15 Procurement Contract	D01 Project Budget	D02 Project Schedule Table	D03 Design Drawing	D04 Procurement Schedule Table	D05 Procurement Budget	D06 Contractor Evaluation Table	D07 List of Certificated Subcontractor	D08 Sub-working Fees	D09 Notification List	D10 Comparison and Negotiation List	D11 Main Contract
A01	Purchasing Schedule	(syn,1,1)		(rt,0.5,1)		(in,0.8,1)										(rt,0.5,1)	(rt,0.5,1)	(in,0.8,1)		implicit (rt*rt) (in*rt) (syn*rt)	implicit (in*rt)	implicit (rt*rt)			implicit (rt*rt)		implicit (rt*rt) (in*rt) (syn*rt)
A02	Design Drawing		(syn,1,1)	(rt,0.5,1)		(rt,0.5,1)												(syn,1,1)		implicit (syn*rt)					implicit (syn*rt)	implicit (in*rt)
A03	Purchasing Specification	(rt,0.5,1)	(rt,0.5,1)	(syn,1,1)		(in,0.8,1)				(rt,0.5,1)							implicit (rt*rt)	implicit (syn*rt)	implicit (rt*rt)	implicit (syn*rt)		implicit (rt*rt)			implicit (in*rt)		
A04	Subcontractor List				(syn,1,1)		(in,0.8,1)	(rt,0.5,1)	(in,0.8,1)	(rt,0.5,1)	(rt,0.5,1)											implicit (rt*rt) (in*rt) (syn*rt)	(in,0.8,1)		implicit (in*rt)	implicit (in*rt)	
A05	Procurement Schedule	(in,0.8,1)	(rt,0.5,1)	(in,0.8,1)			(syn,1,1)	(rt,0.5,1)	(in,0.5,1)				(rt,0.5,1)			(rt,0.5,1)	implicit (in*rt) (syn*rt)	implicit (in*rt) (syn*rt)	implicit (in*rt)	(syn,1,1)	implicit (rt*rt)	implicit (in*rt)			implicit (in*rt)		implicit (syn*rt) (in*rt)
A06	Procurement Request Form				(in,0.8,1)	(rt,0.5,1)	(syn,1,1)	(rt,0.5,1)			(in,0.8,1)		(in,0.8,1)						implicit (syn*rt)			implicit (in*rt)		implicit (rt*rt)			implicit (syn*rt)
A07	Inquiry Sheet				(rt,0.5,1)	(rt,0.5,1)	(rt,0.5,1)	(syn,1,1)			(in,0.8,1)		(in,0.8,1)						implicit (syn*rt)			implicit (in*rt)		implicit (rt*rt)			implicit (syn*rt)
A08	Subcontractor Evaluation Table				(in,0.8,1)				(syn,1,1)	(in,0.8,1)	(in,0.5,1)			(rt,0.5,1)			implicit (syn*rt)		implicit (syn*rt)	implicit (in*rt)	(syn,1,1)	implicit (in,0.5,1)	implicit (rt*rt) (in*rt) (syn*rt)	implicit (in*rt) (syn*rt) (in*rt)		implicit (rt*rt) (syn*rt)	implicit (rt*rt)
A09	Qualified Subcontractor List			(rt,0.5,1)	(rt,0.5,1)				(in,0.8,1)	(syn,1,1)	(in,0.8,1)						implicit (in*rt)	implicit (rt*rt)	implicit (rt*rt)	implicit (rt*rt)		implicit (in,0.8,1)	implicit (rt*rt) (in*rt) (syn*rt)	(in,0.8,1)		implicit (rt*rt) (in*rt) (syn*rt)	implicit (rt*rt) (in*rt)
A10	Supplier Evaluation Table				(rt,0.5,1)				(rt,0.5,1)	(in,0.5,1)	(syn,1,1)	(in,0.5,1)	(rt,0.5,1)								implicit (rt*rt) (in*rt) (syn*rt)	implicit (rt*rt) (in*rt)			implicit (rt*rt)	implicit (in*rt)
A11	Quotation Request						(in,0.8,1)	(in,0.8,1)				(syn,1,1)	(rt,0.5,1)	(rt,0.5,1)	(rt,0.5,1)	(rt,0.5,1)						implicit (in*rt)			implicit (rt*rt)	implicit (syn*rt)	implicit (syn*rt)
A12	Price Evaluation Table				(rt,0.5,1)						(rt,0.5,1)	(syn,1,1)	(in,0.8,1)	(in,0.8,2)	(rt,0.5,1)				implicit (rt*rt)	implicit (syn*rt)	implicit (in*rt)	implicit (rt*rt)		(in,0.8,1)	implicit (rt*rt)	implicit (in*rt) (rt*rt)	implicit (in*rt) (syn*rt)
A13	Comparison and Negotiation List						(in,0.8,1)	(in,0.8,1)	(rt,0.5,1)	(rt,0.5,1)	(in,0.5,1)	(rt,0.5,1)	(rt,0.8,1)	(syn,1,1)	(rt,0.5,1)							implicit (rt*rt) (in*rt) (syn*rt)	implicit (in*rt) (in*rt) (syn*rt)	implicit (rt*rt) (in*rt) (syn*rt)	implicit (syn*rt) (syn,1,1)	implicit (syn*rt)
A14	Detailed Prices Comparison List										(rt,0.5,1)	(in,0.8,1)	(rt,0.5,1)	(syn,1,1)	(rt,0.5,1)							implicit (in*rt)			implicit (rt*rt) (in*rt)	implicit (syn*rt) (syn*rt)
A15	Procurement Contract	(rt,0.5,1)			(rt,0.5,1)						(rt,0.5,1)	(in,0.8,1)	(rt,0.5,1)	(rt,0.5,1)	(syn,1,1)	implicit (rt*rt) (syn*rt)	implicit (in*rt) (syn*rt)		implicit (rt*rt) (syn*rt)	implicit (rt*rt) (syn*rt)		implicit (rt*rt)			implicit (syn*rt)	implicit (syn*rt) (syn,1,1)
D01	Project Budget	(rt,0.5,1)		implicit (rt*rt)	implicit (in*rt) (syn*rt)						implicit (in*rt)					implicit (in*rt) (syn*rt)	(syn,1,1)	(rt,0.5,1)		(rt,0.5,1)	(in,0.8,1)			(rt,0.5,1)			(rt,0.5,1)
D02	Project Schedule Table	(in,0.8,1)		implicit (in*rt)	implicit (in*rt) (syn*rt)				implicit (syn*rt) (in*rt)		implicit (in*rt)					implicit (in*rt) (syn*rt)	(rt,0.5,1)	(syn,1,1)		(in,0.8,1)	(rt,0.5,1)	(in,0.8,1)					(rt,0.5,1)
D03	Design Drawing		(syn,1,1)	implicit (rt*rt)	implicit (in*rt)				implicit (in*rt)		implicit (in*rt)							(syn,1,1)						(rt,0.5,1)	(rt,0.5,1)		
D04	Procurement Schedule Table	implicit (rt*rt) (in*rt) (syn*rt)	implicit (syn*rt)	implicit (syn*rt)		(syn,1,1)	implicit (syn*rt)	implicit (syn*rt)	implicit (in*rt)		implicit (syn*rt)					implicit (syn*rt)	(rt,0.5,1)	(in,0.8,1)		(syn,1,1)	(rt,0.5,2)	(rt,0.5,1)					(rt,0.5,1)
D05	Procurement Budget	implicit (in*rt) (rt*rt)				implicit (in*rt)			implicit (in*rt)		implicit (in*rt)					implicit (in*rt)	(in,0.8,1)	(rt,0.5,1)		(rt,0.5,1)	(syn,1,1)	(rt,0.5,1)		(in,0.8,1)			(rt,0.5,1)
D06	Contractor Evaluation Table	implicit (in*rt)		implicit (in*rt)	implicit (in*rt) (in*rt)	implicit (syn*rt)			(syn,1,1)	(in,0.8,1)	implicit (in*rt)		implicit (in*rt)	implicit (in*rt) (syn*rt)		implicit (syn*rt)		(in,0.8,1)		(rt,0.5,1)	(rt,0.5,1)	(syn,1,1)	(in,0.8,1)	(in,0.8,1)	(in,0.8,1)	(in,0.8,1)	(rt,0.5,1)
D07	List of Certificated Subcontractors				(in,0.8,1)	implicit (in*rt) (rt*rt)	implicit (in*rt)		implicit (in*rt)	implicit (in*rt) (rt*rt)		implicit (rt*rt)	implicit (in*rt)									(in,0.8,1)	(syn,1,1)			(in,0.8,1)	
D08	Sub-working Fees	implicit (rt*rt)	implicit (syn*rt)			implicit (in*rt)			implicit (syn*rt)		implicit (in*rt) (in*rt)		implicit (in*rt) (syn*rt)	implicit (in*rt) (rt*rt)	implicit (in*rt)					(rt,0.5,1)	(rt,0.5,1)	(in,0.8,1)	(in,0.8,1)	(syn,1,1)	(rt,0.5,1)	(rt,0.5,1)	
D09	Notification List		implicit (syn*rt)	implicit (rt*rt)	implicit				implicit (in*rt)	(in,0.8,1)	implicit (in*rt)		implicit (in*rt)	implicit (in*rt)					(rt,0.5,1)		(in,0.8,1)		(rt,0.5,1)	(syn,1,1)	(in,0.8,1)	
D10	Comparison & Negotiation List			implicit (in*rt)		implicit (syn*rt) (in*rt)	implicit (syn*rt)	implicit (in*rt)	implicit (syn*rt)	implicit (syn*rt)	implicit (in*rt)	(syn,1,1)	implicit (syn*rt)	implicit (syn*rt)								(in,0.8,1)	(in,0.8,1)	(rt,0.5,1)	(in,0.8,2)	(syn,1,1)	(rt,0.5,1)
D11	Main Contract	implicit (rt*rt) (syn*rt) (syn*rt)			implicit (rt*rt) (syn*rt)				implicit (syn*rt)	implicit (syn*rt)		implicit (in*rt)	implicit (syn*rt)	implicit (syn*rt)		(syn,1,1)	(rt,0.5,1)	(rt,0.5,1)		(rt,0.5,1)	(in,0.8,1)	(rt,0.5,1)				(rt,0.5,1)	(syn,1,1)

Figure 4.13 Data Thesaurus [Best practice company 3]

It is possible for both explicit and implicit relationships to occur more than once between a given pair of names when several processes are analyzed simultaneously. Therefore, the following two kinds of multiple relationships must be distinguished (Castano and De Antonellis, 1998).

◆ The homogeneous multiple relationship; i.e., when the same type of semantic relationships occur multiple times between two names. An explicit relationship is a homogeneous multiple if its occurrence number $k > 1$. Such is denoted by the symbol "$\mathfrak{R}^k_{hom}$". An implicit relationship is a homogeneous multiple if it involves explicit relationships of the same relationship type and at least one of them is multiple. The symbol "$\Rightarrow^k_{hom}$" is used.

◆ The heterogeneous multiple relationship; i.e., when different types of relationships occur multiple times between two names. An explicit relationship is a heterogeneous multiple if different explicit types exist for the same pair of names. Such is denoted by the symbol "$\mathfrak{R}^k_{het}$". An implicit relationship is a heterogeneous multiple if it involves explicit relationships that are heterogeneous multiples. The symbol "$\Rightarrow^k_{het}$" is used.

Step 2. Create a semantic similarity matrix

The corresponding semantic similarity matrix can be transferred from the created data thesaurus to cluster data entities in the next step. Equation 4.1 ~ 4.4 show the transformation functions corresponding to the relationships within the data thesaurus.

$$A_{hom}(a_{ij}) = a_{ij}^{1/k} \quad \text{..(4.1)}$$

where A_{hom} represents the homogeneous semantic affinity; a_{ij} represents the affinity of relationship of n_i and n_j; and k represents the number of times that the involved relationship has occurred.

48

As demonstrated in Equation 4.1, the greater the number of identified relationship occurrences (k) between two data entities, the higher their total affinity.

$$A_{het}(a_{ij-R_1}, a_{ij-R_2}) = \sqrt{a_{ij-R_1} \cdot a_{ij-R_2}}$$

....(4.2)

where A_{het} represents the heterogeneous semantic affinity; a_{ij-R1} represents the R_1 affinity of relationship of n_i and n_j; and a_{ij-R2} represents the R_2 affinity of relationship of n_i and n_j. R_1, $R_2 \in \{SYN, BT/NT, RT\}$

$$A_{imp}(a_{ih}, a_{hj}) = a_{ih} \cdot a_{hj}$$ ....(4.3)

where A_{imp} represents the affinity function of the implicit relationship; a_{ih} represents the explicit affinity of n_i and n_h; and a_{hj} represents the explicit affinity of n_h and n_j.

Applying Equation 4.1 ~ 4.3, the affinity coefficient of two names n_i and n_j, represented by $AC(n_i, n_j)$, can be evaluated between names in the thesaurus as follows:

$$AC(n_i, n_j) \begin{cases} 1 & if \quad n_i = n_j \\ A_{hom}(a_{ij}) & if \quad n_i \Re^k_{hom} n_j, \ k \geq 1 \\ A_{het}(A_{hom}(a_{ij-R_1}), A_{hom}(a_{ij-R_2})) & if \quad n_i \Re_{het} n_j \\ A_{hom}(A_{imp}(a_{ih}, a_{hj})) & if \quad n_i \Rightarrow^k_{hom} n_j \\ A_{het}(A_{imp}(a_{ih}, a_{hj}), A_{imp}(a_{ip}, a_{pj})) & if \quad n_i \Rightarrow_{het} n_j \\ 0 & otherwise \end{cases}$$ ....(4.4)

where $AC(n_i, n_j)$ represents a numerical value in the range [0,1] and a_{ij-R1} (respectively, a_{ij-R2}) represents the affinity of n_i and n_j corresponding to the R_1 (respectively, R_2) relationship; i.e., R_1, $R_2 \in \{SYN, BT/NT, RT\}$.

Applying Equation 4.4, a semantic similarity matrix can be created based on the data thesaurus. Figure 4.14, Figure 4.15 and Figure 4.16 show an example data semantic similarity matrix derived from Figure 4.11, Figure 4.12 and Figure 4.13.

Ni \ Nj			Benchmarking Company															Best-practice Company (1)																	
			A01	A02	A03	A04	A05	A06	A07	A08	A09	A10	A11	A12	A13	A14	A15	B01	B02	B03	B04	B05	B06	B07	B08	B09	B10	B11	B12	B13	B14	B15	B16	B17	B18
		Iy	Purchasing Schedule	Design Drawing	Purchasing Specification	Subcontractor List	Procurement Schedule	Procurement Request Form	Inquiry Sheet	Subcontractor Evaluation Table	Qualified Subcontractor List	Supplier Evaluation Table	Quotation Request	Price Evaluation Table	Comparison and Negotiation List	Detailed Prices Comparison List	Procurement Contract	Procurement Schedule Table	Design Task Table	List of Certificated Contractor	Check List of Construction Progress	Revised Procurement Schedule Table	Inquiry Sheet	List of Qualified Subcontractors	Subcontractor Evaluation Table	Bill of Inquiries	Quotation Request	List of Procurement Quotation	Bill of Quotation Request	Bill of Comparison and Negotiation	Joint Venture Cooperation	Bill Collation	Notification List for Modifying	Main Contract	Contract Duplication
Benchmarking Company	A01	Purchasing Schedule	1.000	0.000	0.500	0.000	0.800	0.000	0.000	0.000	0.000	0.000	0.000	0.000	0.000	0.000	0.500	0.500	0.250	0.000	0.250	0.500	0.000	0.000	0.000	0.000	0.250	0.000	0.000	0.000	0.000	0.000	0.000	0.500	0.707
	A02	Design Drawing	0.000	1.000	0.500	0.000	0.500	0.000	0.000	0.000	0.000	0.000	0.000	0.000	0.000	0.000	0.000	0.500	1.000	0.000	0.000	0.447	0.500	0.000	0.000	0.000	0.000	0.500	0.500	0.000	0.000	0.000	0.000	0.000	0.000
	A03	Purchasing Specification	0.500	0.500	1.000	0.000	0.800	0.000	0.000	0.000	0.500	0.000	0.000	0.000	0.000	0.000	0.000	0.447	0.500	0.000	0.000	0.400	0.000	0.500	0.000	0.000	0.000	0.000	0.000	0.000	0.000	0.000	0.000	0.000	0.000
	A04	Subcontractor List	0.800	0.000	0.000	1.000	0.000	0.800	0.500	0.800	0.500	0.500	0.000	0.000	0.000	0.000	0.000	0.000	0.000	0.800	0.000	0.000	0.500	0.800	0.602	0.400	0.400	0.000	0.400	0.000	0.400	0.000	0.000	0.400	0.400
	A05	Procurement Schedule	0.000	0.500	0.800	0.000	1.000	0.500	0.500	0.000	0.000	0.000	0.000	0.500	0.000	0.000	0.500	1.000	0.566	0.000	0.400	0.800	0.500	0.000	0.000	0.400	0.400	0.000	0.000	0.250	0.000	0.000	0.000	0.532	0.447
	A06	Procurement Request Form	0.000	0.000	0.000	0.800	0.500	1.000	0.500	0.000	0.000	0.000	0.800	0.000	0.800	0.000	0.000	0.500	0.000	0.640	0.000	0.400	0.500	0.640	0.000	0.400	0.800	0.000	0.640	0.800	0.000	0.400	0.000	0.000	0.000
	A07	Inquiry Sheet	0.000	0.000	0.000	0.500	0.500	0.500	1.000	0.000	0.000	0.000	0.800	0.000	0.800	0.000	0.000	0.500	0.500	0.400	0.000	0.400	1.000	0.400	0.000	0.800	0.757	0.447	0.589	0.800	0.000	0.632	0.400	0.000	0.000
	A08	Subcontractor Evaluation Table	0.000	0.000	0.000	0.800	0.000	0.000	0.000	1.000	0.800	0.500	0.000	0.000	0.500	0.000	0.000	0.000	0.000	0.716	0.500	0.000	0.000	0.743	1.000	0.000	0.000	0.000	0.000	0.707	0.500	0.250	0.000	0.500	0.500
	A09	Qualified Subcontractor List	0.000	0.000	0.500	0.500	0.000	0.000	0.000	0.800	1.000	0.500	0.000	0.000	0.500	0.000	0.000	0.000	0.000	0.566	0.000	0.000	0.000	1.000	0.635	0.000	0.500	0.000	0.500	0.500	0.500	0.250	0.000	0.500	0.500
	A10	Supplier Evaluation Table	0.000	0.000	0.000	0.500	0.000	0.000	0.000	0.500	0.500	1.000	0.500	0.000	0.500	0.000	0.000	0.000	0.000	0.506	0.400	0.000	0.000	0.504	0.800	0.000	0.500	0.000	0.500	0.447	0.400	0.250	0.000	0.400	0.400
	A11	Quotation Request	0.000	0.000	0.000	0.000	0.000	0.800	0.800	0.000	0.000	0.500	1.000	0.500	0.500	0.500	0.500	0.000	0.400	0.500	0.000	0.500	0.598	0.447	0.400	0.689	1.000	0.447	0.800	0.292	0.400	0.368	0.316	0.447	0.400
	A12	Price Evaluation Table	0.000	0.000	0.000	0.000	0.500	0.000	0.000	0.000	0.000	0.000	0.500	1.000	0.800	0.800	0.500	0.500	0.000	0.000	0.000	0.400	0.000	0.000	0.250	0.000	0.500	0.354	0.316	0.500	0.250	0.316	0.316	0.354	0.250
	A13	Comparison and Negotiation List	0.000	0.000	0.000	0.000	0.000	0.800	0.800	0.500	0.500	0.500	0.500	0.800	1.000	0.500	0.500	0.000	0.000	0.000	0.250	0.000	0.600	0.500	0.532	0.400	0.354	0.420	0.368	1.000	0.353	0.500	0.500	0.420	0.353
	A14	Detailed Prices Comparison List	0.000	0.000	0.000	0.000	0.000	0.000	0.000	0.000	0.000	0.000	0.500	0.800	0.500	1.000	0.500	0.000	0.000	0.000	0.250	0.000	0.000	0.000	0.000	0.250	0.500	0.250	0.316	0.368	0.000	0.500	0.500	0.354	0.250
	A15	Procurement Contract	0.500	0.000	0.000	0.000	0.500	0.000	0.000	0.000	0.000	0.000	0.500	0.500	0.500	0.500	1.000	0.420	0.000	0.000	0.500	0.368	0.000	0.500	0.500	0.000	0.500	0.500	0.447	0.420	0.500	0.354	0.354	1.000	0.800
Best-practice Company (1)	B01	Procurement Schedule Table	0.500	0.500	0.447	0.000	1.000	0.500	0.500	0.000	0.000	0.000	0.000	0.500	0.000	0.000	0.420	1.000	0.000	0.000	0.000	0.800	0.000	0.000	0.000	0.000	0.000	0.000	0.000	0.000	0.000	0.000	0.000	0.500	0.500
	B02	Design Task Table	0.250	1.000	0.500	0.000	0.566	0.000	0.500	0.000	0.000	0.000	0.400	0.000	0.000	0.000	0.000	0.000	1.000	0.000	0.000	0.500	0.500	0.000	0.000	0.000	0.000	0.500	0.500	0.000	0.000	0.000	0.000	0.000	0.000
	B03	List of Certificated Contractor	0.000	0.000	0.000	0.800	0.000	0.640	0.400	0.716	0.566	0.506	0.500	0.000	0.000	0.000	0.000	0.000	0.000	1.000	0.000	0.000	0.000	0.800	0.800	0.000	0.500	0.000	0.000	0.000	0.000	0.000	0.000	0.000	0.000
	B04	Check List of Construction Progress	0.250	0.000	0.000	0.000	0.400	0.000	0.000	0.500	0.000	0.400	0.000	0.000	0.250	0.250	0.500	0.000	0.000	0.000	1.000	0.500	0.000	0.000	0.500	0.000	0.000	0.000	0.000	0.000	0.500	0.500	0.500	0.500	0.500
	B05	Revised Procurement Schedule Table	0.500	0.447	0.400	0.000	0.800	0.400	0.400	0.000	0.000	0.000	0.500	0.400	0.000	0.000	0.368	0.800	0.500	0.000	0.500	1.000	0.000	0.000	0.000	0.000	0.500	0.000	0.000	0.000	0.000	0.000	0.000	0.500	0.500
	B06	Inquiry Sheet	0.000	0.500	0.000	0.500	0.500	0.500	1.000	0.000	0.000	0.000	0.598	0.000	0.800	0.000	0.000	0.000	0.500	0.000	0.000	0.000	1.000	0.000	0.000	0.800	0.800	0.500	0.500	0.000	0.000	0.000	0.000	0.000	0.000
	B07	List of Qualified Subcontractors	0.000	0.000	0.500	0.800	0.000	0.640	0.400	0.743	1.000	0.504	0.447	0.000	0.500	0.000	0.500	0.000	0.000	0.800	0.000	0.000	0.000	1.000	0.800	0.000	0.500	0.000	0.500	0.000	0.500	0.000	0.000	0.500	0.500
	B08	Subcontractor Evaluation Table	0.000	0.000	0.000	0.602	0.000	0.000	0.000	1.000	0.635	0.800	0.400	0.250	0.532	0.000	0.500	0.000	0.000	0.800	0.500	0.000	0.000	0.800	1.000	0.000	0.000	0.000	0.000	0.500	0.500	0.000	0.000	0.500	0.500
	B09	Bill of Inquiries	0.000	0.000	0.000	0.400	0.400	0.400	0.800	0.000	0.000	0.000	0.689	0.000	0.400	0.250	0.000	0.000	0.000	0.000	0.000	0.000	0.800	0.000	0.000	1.000	0.800	0.500	0.800	0.000	0.000	0.500	0.500	0.000	0.000
	B10	Quotation Request	0.250	0.000	0.000	0.400	0.400	0.800	0.757	0.000	0.500	0.500	1.000	0.500	0.354	0.500	0.500	0.000	0.000	0.500	0.000	0.500	0.800	0.500	0.000	0.800	1.000	0.500	0.800	0.000	0.000	0.500	0.000	0.000	0.000
	B11	List of Procurement Quotation	0.000	0.500	0.000	0.000	0.000	0.000	0.447	0.000	0.000	0.000	0.447	0.354	0.420	0.250	0.500	0.000	0.500	0.000	0.000	0.000	0.500	0.000	0.000	0.500	0.500	1.000	0.500	0.707	0.500	0.500	0.500	0.500	0.500
	B12	Bill of Quotation Request	0.000	0.500	0.000	0.400	0.000	0.640	0.589	0.000	0.500	0.500	0.800	0.316	0.368	0.316	0.447	0.000	0.500	0.000	0.000	0.000	0.500	0.500	0.000	0.800	0.800	0.500	1.000	0.500	0.500	0.500	0.500	0.500	0.500
	B13	Bill of Comparison and Negotiation	0.000	0.000	0.000	0.000	0.250	0.800	0.800	0.707	0.500	0.447	0.292	0.500	1.000	0.368	0.420	0.000	0.000	0.000	0.000	0.000	0.000	0.000	0.500	0.000	0.000	0.707	0.500	1.000	0.500	0.500	0.500	0.500	0.500
	B14	Joint Venture Cooperation	0.000	0.000	0.000	0.400	0.000	0.000	0.000	0.500	0.500	0.400	0.400	0.250	0.353	0.000	0.500	0.000	0.000	0.000	0.500	0.000	0.000	0.500	0.500	0.000	0.000	0.500	0.500	0.500	1.000	0.500	0.000	0.500	0.500
	B15	Bill Collation	0.000	0.000	0.000	0.000	0.000	0.400	0.632	0.250	0.250	0.250	0.368	0.316	0.500	0.500	0.354	0.000	0.000	0.000	0.500	0.000	0.000	0.000	0.000	0.500	0.500	0.500	0.500	0.500	0.500	1.000	0.500	0.500	0.500
	B16	Notification List for Modifying	0.000	0.000	0.000	0.000	0.000	0.000	0.400	0.000	0.000	0.000	0.316	0.316	0.500	0.500	0.354	0.000	0.000	0.000	0.500	0.000	0.000	0.000	0.000	0.500	0.000	0.500	0.500	0.500	0.000	0.500	1.000	0.500	0.500
	B17	Main Contract	0.500	0.000	0.000	0.400	0.532	0.000	0.000	0.500	0.500	0.400	0.447	0.354	0.420	0.354	1.000	0.500	0.000	0.000	0.500	0.500	0.000	0.500	0.500	0.000	0.000	0.500	0.500	0.500	0.500	0.500	0.500	1.000	0.800
	B18	Contract Duplication	0.707	0.000	0.000	0.400	0.447	0.000	0.000	0.500	0.500	0.400	0.400	0.250	0.353	0.250	0.800	0.500	0.000	0.000	0.500	0.500	0.000	0.500	0.500	0.000	0.000	0.500	0.500	0.500	0.500	0.500	0.500	0.800	1.000

Figure 4.14 Data Semantic Similarity Matrix [Best practice company 1] (derived from Figure 4.11)

Data Semantic Similarity Matrix table with columns grouped under "Benchmarking Company" (A01–A15) and "Best-practice Company (2)" (C01–C16), and rows grouped under "Benchmarking Company" (A01–A15) and "Best-practice Company (2)" (C01–C16).

Column headers (Nj):
- A01 Purchasing Schedule
- A02 Design Drawing
- A03 Purchasing Specification
- A04 Subcontractor List
- A05 Procurement Schedule
- A06 Procurement Request Form
- A07 Inquiry Sheet
- A08 Subcontractor Evaluation Table
- A09 Qualified Subcontractor List
- A10 Supplier Evaluation Table
- A11 Quotation Request
- A12 Price Evaluation Table
- A13 Comparison and Negotiation List
- A14 Detailed Prices Comparison List
- A15 Procurement Contract
- C01 Project Schedule Table
- C02 Project Budget
- C03 Project Process Method
- C04 Subcontractor List
- C05 Subcontractor Labors Activity
- C06 Financial Statement
- C07 Procurement Schedule Table
- C08 Procurement Request Form
- C09 Inquiry Sheet
- C10 List of Certificated Subcontractors
- C11 Qualified Subcontractor List
- C12 Inquiries List
- C13 Subcontractor Quotation List
- C14 Comparison and Negotiation List
- C15 Detailed Prices List
- C16 Procurement Contract

Row labels (Ni) mirror the same set: A01–A15 (Benchmarking Company) and C01–C16 (Best-practice Company (2)).

Figure 4.15 Data Semantic Similarity Matrix [Best practice company 2]
(derived from Figure 4.12)

51

Ni \ Nj (ly)		Benchmarking Company															Best-practice Company (3)										
		A01 Purchasing Schedule	A02 Design Drawing	A03 Purchasing Specification	A04 Subcontractor List	A05 Procurement Schedule	A06 Procurement Request Form	A07 Inquiry Sheet	A08 Subcontractor Evaluation Table	A09 Qualified Subcontractor List	A10 Supplier Evaluation Table	A11 Quotation Request	A12 Price Evaluation Table	A13 Comparison and Negotiation List	A14 Detailed Prices Comparison List	A15 Procurement Contract	D01 Project Budget	D02 Project Schedule Table	D03 Design Drawing	D04 Procurement Schedule Table	D05 Procurement Budget	D06 Contractor Evaluation Table	D07 List of Certificated Subcontractor	D08 Sub-working Fees	D09 Notification List	D10 Comparison and Negotiation List	D11 Main Contract
Benchmarking Company	A01 Purchasing Schedule	1.000	0.000	0.500	0.000	0.800	0.000	0.000	0.000	0.000	0.000	0.000	0.000	0.000	0.000	0.500	0.500	0.800	0.000	0.504	0.400	0.640	0.000	0.250	0.000	0.000	0.368
	A02 Design Drawing	0.000	1.000	0.500	0.000	0.500	0.000	0.000	0.000	0.000	0.000	0.000	0.000	0.000	0.000	0.000	0.000	0.000	1.000	0.500	0.000	0.000	0.000	0.500	0.500	0.000	0.000
	A03 Purchasing Specification	0.500	0.500	1.000	0.000	0.800	0.000	0.000	0.000	0.500	0.000	0.000	0.000	0.000	0.000	0.000	0.250	0.400	0.500	0.800	0.000	0.400	0.000	0.000	0.400	0.000	0.000
	A04 Subcontractor List	0.800	0.000	0.000	1.000	0.000	0.800	0.500	0.800	0.500	0.500	0.000	0.000	0.000	0.000	0.000	0.000	0.000	0.000	0.000	0.000	0.589	0.800	0.000	0.400	0.640	0.000
	A05 Procurement Schedule	0.000	0.500	0.800	0.000	1.000	0.500	0.500	0.000	0.000	0.000	0.000	0.500	0.000	0.000	0.500	0.447	0.716	0.500	1.000	0.500	0.500	0.000	0.400	0.000	0.000	0.707
	A06 Procurement Request Form	0.000	0.000	0.000	0.800	0.500	1.000	0.500	0.000	0.000	0.000	0.800	0.000	0.800	0.000	0.000	0.000	0.000	0.000	0.500	0.000	0.000	0.640	0.000	0.000	0.800	0.000
	A07 Inquiry Sheet	0.000	0.000	0.000	0.500	0.500	0.500	1.000	0.000	0.000	0.000	0.800	0.000	0.800	0.000	0.000	0.000	0.000	0.000	0.500	0.000	0.000	0.400	0.000	0.000	0.800	0.000
	A08 Subcontractor Evaluation Table	0.000	0.000	0.000	0.800	0.000	0.000	0.000	1.000	0.800	0.500	0.000	0.000	0.500	0.000	0.000	0.000	0.800	0.000	0.500	0.500	1.000	0.500	0.800	0.716	0.543	0.500
	A09 Qualified Subcontractor List	0.000	0.000	0.500	0.500	0.000	0.000	0.000	0.800	1.000	0.500	0.000	0.000	0.500	0.000	0.000	0.000	0.640	0.400	0.400	0.400	0.800	0.636	0.506	0.800	0.589	0.400
	A10 Supplier Evaluation Table	0.000	0.000	0.000	0.500	0.000	0.000	0.000	0.500	0.500	1.000	0.500	0.000	0.500	0.000	0.000	0.000	0.000	0.000	0.000	0.000	0.447	0.316	0.000	0.400	0.500	0.000
	A11 Quotation Request	0.000	0.000	0.000	0.000	0.000	0.800	0.800	0.000	0.000	0.500	1.000	0.500	0.500	0.500	0.500	0.000	0.000	0.000	0.000	0.000	0.000	0.000	0.400	0.000	0.500	0.500
	A12 Price Evaluation Table	0.000	0.000	0.000	0.000	0.500	0.000	0.000	0.000	0.000	0.000	0.500	1.000	0.800	0.800	0.500	0.400	0.000	0.400	0.500	0.640	0.640	0.000	0.800	0.400	0.566	0.500
	A13 Comparison and Negotiation List	0.000	0.000	0.000	0.000	0.000	0.800	0.800	0.500	0.500	0.500	0.500	0.800	1.000	0.500	0.500	0.000	0.000	0.000	0.000	0.000	0.543	0.447	0.566	0.566	1.000	0.707
	A14 Detailed Prices Comparison List	0.000	0.000	0.000	0.000	0.000	0.000	0.000	0.000	0.000	0.000	0.500	0.800	0.500	1.000	0.500	0.000	0.000	0.000	0.000	0.000	0.000	0.000	0.640	0.000	0.500	0.500
	A15 Procurement Contract	0.500	0.000	0.000	0.000	0.500	0.000	0.000	0.000	0.000	0.000	0.500	0.500	0.500	0.500	1.000	0.353	0.447	0.000	0.707	0.500	0.500	0.000	0.400	0.000	0.707	1.000
Best-practice Company (3)	D01 Project Budget	0.500	0.000	0.250	0.000	0.447	0.000	0.000	0.000	0.000	0.000	0.000	0.400	0.000	0.000	0.353	1.000	0.500	0.000	0.500	0.800	0.000	0.000	0.500	0.000	0.000	0.500
	D02 Project Schedule Table	0.800	0.000	0.400	0.000	0.716	0.000	0.000	0.800	0.640	0.000	0.000	0.000	0.000	0.000	0.447	0.500	1.000	0.000	0.800	0.500	0.800	0.000	0.000	0.000	0.000	0.500
	D03 Design Drawing	0.000	1.000	0.500	0.000	0.500	0.000	0.000	0.000	0.400	0.000	0.000	0.400	0.000	0.000	0.000	0.000	0.000	1.000	0.000	0.000	0.000	0.000	0.500	0.500	0.000	0.000
	D04 Procurement Schedule Table	0.504	0.500	0.800	0.000	1.000	0.500	0.500	0.500	0.400	0.000	0.000	0.500	0.000	0.000	0.707	0.500	0.800	0.000	1.000	0.500	0.500	0.000	0.000	0.000	0.000	0.500
	D05 Procurement Budget	0.400	0.000	0.000	0.000	0.500	0.000	0.000	0.500	0.400	0.000	0.000	0.640	0.000	0.000	0.500	0.800	0.500	0.000	0.500	1.000	0.500	0.000	0.800	0.000	0.000	0.500
	D06 Contractor Evaluation Table	0.640	0.000	0.400	0.589	0.500	0.000	0.000	1.000	0.800	0.447	0.000	0.640	0.543	0.000	0.500	0.000	0.800	0.000	0.500	0.500	1.000	0.800	0.800	0.800	0.800	0.500
	D07 List of Certificated Subcontractors	0.000	0.000	0.000	0.800	0.000	0.640	0.400	0.500	0.636	0.316	0.000	0.000	0.447	0.000	0.000	0.000	0.000	0.000	0.000	0.000	0.800	1.000	0.000	0.000	0.800	0.000
	D08 Sub-working Fees	0.250	0.500	0.000	0.000	0.400	0.000	0.000	0.800	0.506	0.000	0.400	0.800	0.566	0.640	0.400	0.500	0.000	0.500	0.000	0.800	0.800	0.000	1.000	0.500	0.500	0.000
	D09 Notification List	0.000	0.500	0.400	0.400	0.000	0.000	0.000	0.716	0.800	0.400	0.000	0.400	0.566	0.000	0.000	0.000	0.000	0.500	0.000	0.000	0.800	0.000	0.500	1.000	0.800	0.000
	D10 Comparison & Negotiation List	0.000	0.000	0.000	0.640	0.000	0.800	0.800	0.543	0.589	0.500	0.500	0.566	1.000	0.500	0.707	0.000	0.000	0.000	0.000	0.000	0.800	0.800	0.500	0.800	1.000	0.500
	D11 Main Contract	0.368	0.000	0.000	0.000	0.707	0.000	0.000	0.500	0.400	0.000	0.500	0.500	0.707	0.500	1.000	0.500	0.500	0.000	0.500	0.500	0.500	0.000	0.000	0.000	0.500	1.000

Figure 4.16 Data Semantic Similarity Matrix [Best practice company 3]
(derived from Figure 4.13)

Continuing with the previous example, because only one implicit relationship (BT,RT) exists between "A07" and "B03" data entities (i.e., no other different relationship exists between these two entities), the condition for calculating affinity coefficient is determined as $n_{B03} \Rightarrow_{\text{hom}} n_{A07}$, where n_{A07} = "A07" data entity and n_{B03} = "B03" data entity in the processes. Accordingly,

$$AC(n_{A07}, n_{B03}) = A_{\text{hom}} (A_{\text{imp}} (\alpha_{A07-A04}, \alpha_{A04-B03}))$$

$$= a_{A07-A04} \times a_{A04-B03j} = a_{BT} \times a_{RT} = 0.8 \times 0.5$$

$$= 0.4$$

Step 3. Data semantic hierarchy creation.

Based on the semantic similarity matrix, the data semantic hierarchy can be created using clustering techniques. Hierarchical agglomerative clustering techniques that may be applied include the single linkage agglomerative, complete linkage agglomerative, weighted / unweighted arithmetic average clustering techniques, variances in the distance function of each clustering method produces clustering techniques to obtain different cluster results, which were then summarized to develop the semantic hierarchy (Tsai, 2007). Figure 4.17 shows the data semantic hierarchy of procurement process from the case study, derived from Figure 4.14-4.16 for the implementation of BOPR.

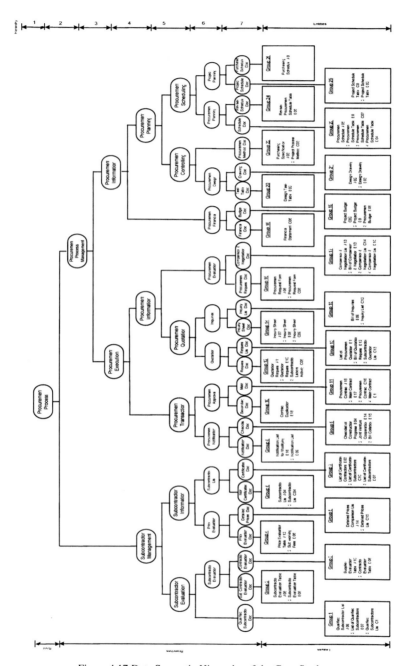

Figure 4.17 Data Semantic Hierarchy of the Case Study

A semantic hierarchy is composed of a root, branches, and leaves. The root describes the abstract concept of the hierarchy. Branches (concept links) present all sub-concepts derived from the root concept, while concept link level presented in each branch shows the hierarchical depth of each sub-concept. Leaves describe data entity clusters within process models. Each leaf corresponds to a specific concept link located at a specific hierarchy level. The higher the level, the more general the semantic concept represented. Therefore, the semantic similarity of two entities can be calculated based on the semantic similarity of their common concept link. Consequently, based on Equation 2.1 in the chapter 2, this study applied Equation 4.5 to calculate the similarity of two entities.

$$SS_{ij} = \delta^{l_{ij}} \dots\dots\dots\dots\dots\dots\dots\dots\dots\dots\dots\dots\dots\dots\dots\dots\dots\dots\dots(4.5)$$

where SS_{ij} represents the semantic similarity of n_i and n_j; δ represents the initial similarity value for bottom level concept links; and l_{ij} represents the hierarchical length of the common link of n_i and n_j.

For data semantic hierarchy, the initial value of δ for bottom level concept links is set as 0.8 in accordance with the experimentation of Castano and De Antonellis (1998), and, because case study data semantic hierarchy has five levels, the data similarity of two data entities is evaluated based on Equation 4.6

$$SS_{ij} = 0.8^{l_{ij}} \; ; \; l_{ij} = l_D = 7 - L_{CL\text{-}ij} + 1\dots\dots\dots\dots\dots\dots\dots\dots\dots\dots\dots\dots\dots\dots(4.6)$$

where l_D represents hierarchy length in the data semantic hierarchy and $L_{CL\text{-}ij}$ represents the hierarchy level of the common concept link of n_i and n_j.

Taking the data entities in the Figure 4.17 as an example, the revised procurement schedule table within the cluster Group 24 and the purchasing schedule within the cluster Group 26 shares a common concept link, "Procurement Scheduling," at hierarchy level 5. Thus, the semantic similarity of the revised procurement schedule table and purchasing schedule can be calculated as $0.8^{(7-5+1)} = 0.512$.

4.2.1.2   Activity semantic hierarchy creation

Calculating the functional similarity of two activities or processes, not only a data semantic hierarchy need to be created, but an activity hierarchy must also be established. Unlike the data semantic hierarchy, selected process activities should be classified into general categories according to their essential functional concepts based on the operational concepts outlined by De Antonellis and Zonta (1990). Only functions acting on resources, i.e., "Creation," "Exchange," "Transformation," "Modification" and "Deletion," are considered in this study. Table 4.1 shows the definitions of these five functional types. The activity semantic hierarchy can be created according to Table 4.1. Figure 4.18 shows the activity semantic hierarchy of procurement process from the case study.

Table 4.1 Definitions of Activity Functions (De Antonellis and Zonta, 1990)

Function Type	Definition	Common Verbs
Creation	Creation of resources as operation output.	compile, construct, define, draw, institute, make, prepare, produce, etc..
Exchange	Exchange of resources with the outside of the task to which the operation belongs.	communicate, deliver, display, distribute, give, obtain, present, receive, show, ask, inform, etc..
Transformation	Transformation of input resources into output resources with a new identity.	compact, compose, cut, decompose, divide, join, merge, split, etc..
Modification/Obser vation	Modification/Observation of the status of input resources, while preserving their identity in the output. The modification can be of content, classification, addition/ subtraction of properties.	check, choose, dismiss, examine, fill in, protect, retrieve, select, sign, test, update, etc..
Deletion	Deletion or removal of resources.	cancel, delete, eliminate, erase, throw away, undo, etc..

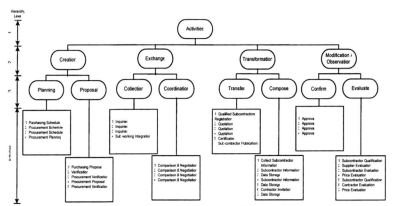

Figure 4.18 Activity Semantic Hierarchy of Procurement Process

Similar to the data similarity calculation, similarities of activity semantic hierarchy are calculated based on Equation 4.5. However, since the activity semantic hierarchy has three levels, the similarity of the two activities is evaluated using Equation 4.7, in which the initial value of δ for activity semantic hierarchy bottom-level concept links is set as 0.5.

$$SS_{ij} = 0.5^{l_i} ; l_{ij} = l_A = 3 - L_{CL-ij} + 1 \dots\dots\dots\dots\dots\dots\dots\dots\dots\dots\dots\dots\dots\dots\dots\dots\dots(4.7)$$

where l_A represents the hierarchy length in the activity semantic hierarchy.

4.2.2 Similarity Analysis

Factors considered in calculating process similarity when creating collaborative processes include: (1) name affinity, (2) name set affinity, (3) information similarity, (4) activity similarity and (5) function similarity. Information, activity and function similarities can be calculated based on name and name set affinities in order to identify overlapping and coupling relationships in selected processes and generate the reengineering measures in terms of "process unification", "activity fusion" or "as usual".

◆ Name affinity:

Based on Equation 4.5, the name affinity of any two entities, represented by "Na", can be determined according to Equation 4.8.

$$Na(n_i, n_j) = \begin{cases} 1 & if\ (n_i = n_j) \\ SS_{ij} & if\ (n_i \neq n_j) \wedge (n_i, n_j \in CL_{ij}) \\ 0 & otherwise \end{cases} \dots\dots\dots\dots\dots\dots\dots\dots\dots\dots\dots(4.8)$$

where $Na(n_i, n_j)$ represents the name affinity of entities n_i and n_j; SS_{ij} represents the semantic similarity of n_i and n_j; and CL_{ij} represents the common concept link of n_i and n_j in the data semantic hierarchy or in the activity semantic hierarchy.

◆ Name set affinity:

Equation 4.8 can be applied to determine the similarity of two names only. However, because processes are aggregations of activities and data entities, the name set affinity parameter denoted by $Nsa(X,Y)$ must be addressed to evaluate the semantic similarity of two sets, such as process input or output data sets. Equation 4.9 shows the name set affinity function based on *name affinity*.

$$Nsa(X,Y) = \frac{\sum_{i=1}^{|X|} \underset{j=1 \to |Y|}{Max}(Na(n_i,n_j)) + \sum_{j=1}^{|Y|} \underset{i=1 \to |X|}{Max}(Na(n_i,n_j))}{|X|+|Y|} \qquad \forall n_i \in X ; \forall n_j \in Y \quad ....(4.9)$$

Where $0 \leq Nsa \leq 1$; X and Y represent two name sets, and $|X|$ and $|Y|$ represent numbers of elements in, respectively, X and Y.

By calculating all pairs of name entities belonging to two sets, the value of $Nsa(X,Y)$ expresses the similarity of two name sets.

Based on Equation 4.8 and 4.9, three functions were developed to evaluate activity and process similarities. These included the (1) Process Information Similarity, (2) Activity Similarity, and (3) Process Functional Similarity functions

4.2.2.1 Process Information Similarity

The process information similarity denoted by $PISim(P_i,P_j)$ is the measure of similarity of input and output information sets corresponding severally to two analyzed processes P_i and P_j. Equation 4.10 shows the function of process information similarity.

$$PISim(P_i,P_j) = \Sigma\ A(P_if_2, P_jf_3)...(4.10)$$

where and $0 \leq PISim(P_i,P_j) \leq 2$. P_i and P_j are respectively the semantic process model of *process i* and of *process j*; f_2 indicates the input data set of a process, and f_3 indicates the output data set of a process.

59

Table 4.2 and Table 4.3 show the name set affinity of input data sets and output data sets of procurement process from benchmarking company (P_1), and best practice company 1 (P_2) of the case study, the $A(P_1.f_2, P_2.f_2)$ equals 0.540. Similarly, the output data set affinity parameters, $A(P_1.f_3, P_2.f_3)$ was also calculated that equals to 0.680. Accordingly, the $PISim(P_1, P_2) = 0.540 + 0.680 = 1.22$.

Table 4.2 Name Set Affinity (Nsa) of Input Data Set of Procurement Process between Benchmarking Company and Best practice Company 1

		Best-practice Company (1)				Maximum of Row
		Procurement Schedule Table (B01)	Design Task Table (B02)	List of Certificated Contractor (B03)	Checklist of Construction Progress (B04)	
Benchmarking Company	Purchasing Schedule (A01)	0.512	0.410	0.210	0.260	0.512
	Design Drawing (A02)	0.410	0.640	0.210	0.260	0.640
	Purchasing Specification (A03)	0.410	0.512	0.210	0.260	0.512
	Subcontractor List (A04)	0.210	0.210	0.640	0.210	0.640
Maximum of Column		0.512	0.640	0.640	0.260	
Sum of Max. of Row + Sum of Max. of Column						4.36
Nsa(P1(input set), P2(input set))						0.54

Table 4.3 Name Set Affinity (Nsa) of Output Data Set of Procurement Process between Benchmarking Company and Best practice Company 1

	Revised Procurement Schedule Table (B05)	Inquiry Sheet (B06)	List of Qualified Subcontractor (B07)	Subcontractor Evaluation Table (B08)	Bill of Inquiries (B09)	Quotation Request (B10)	List of Procurement Quotation (B11)	Bill of Quotation Request (B12)	Bill of Comparison and Negotiation (B13)	Joint Venture Cooperation (B14)	Bill Collation (B15)	Notification List for Modifying (B16)	Main Contract (B17)	Contract Duplication (B18)	Maximum of Row
Procurement Schedule (A05)	0.640	0.260	0.210	0.210	0.260	0.260	0.260	0.260	0.260	0.260	0.260	0.260	0.260	0.260	0.640
Procurement Request Form (A06)	0.260	0.410	0.210	0.210	0.410	0.410	0.410	0.410	0.512	0.330	0.330	0.330	0.330	0.330	0.512
Inquiry Sheet (A07)	0.260	0.800	0.210	0.210	0.640	0.512	0.512	0.512	0.410	0.330	0.330	0.330	0.330	0.330	0.800
Subcontractor Evaluation Table (A08)	0.210	0.210	0.512	0.800	0.210	0.210	0.210	0.210	0.210	0.210	0.210	0.210	0.210	0.210	0.800
Qualified Subcontractors List (A09)	0.210	0.210	0.800	0.512	0.210	0.210	0.210	0.210	0.210	0.210	0.210	0.210	0.210	0.210	0.800
Supplier Evaluation Table (A10)	0.210	0.210	0.512	0.640	0.210	0.210	0.210	0.210	0.210	0.210	0.210	0.210	0.210	0.210	0.640
Quotation Request (A11)	0.260	0.512	0.210	0.210	0.512	0.800	0.640	0.640	0.410	0.330	0.330	0.330	0.330	0.330	0.800
Price Evaluation Table (A12)	0.210	0.210	0.410	0.410	0.210	0.210	0.210	0.210	0.210	0.210	0.210	0.210	0.210	0.210	0.410
Comparison and Negotiation List (A13)	0.260	0.410	0.210	0.210	0.410	0.410	0.410	0.410	0.800	0.330	0.330	0.330	0.330	0.330	0.800
Detailed Prices Comparison List (A14)	0.210	0.210	0.410	0.410	0.210	0.210	0.210	0.210	0.210	0.210	0.210	0.210	0.210	0.210	0.410
Procurement Contract (A15)	0.260	0.330	0.210	0.210	0.330	0.330	0.330	0.330	0.330	0.512	0.512	0.512	0.800	0.640	0.800
Maximum of Column	0.640	0.800	0.800	0.800	0.640	0.800	0.640	0.640	0.800	0.512	0.512	0.512	0.800	0.640	
Sum of Max. of Row + Sum of Max. of Column															16.947
Nsa(P1(output set), P2(output set))															0.68

Table 4.4 and Table 4.5 show the name set affinity of input data sets and output data sets of procurement process from benchmarking company (P_1), and best practice company 2 (P_3) of the case study, the $A(P_1.f_2, P_3.f_2)$ equals 0.590. Similarly, the output data set affinity parameters, $A(P_1.f_3, P_3.f_3)$ was also calculated that equals to 0.710. Accordingly, the $PISim(P_1, P_3) = 0.590 + 0.710 = 1.30$.

Table 4.4 Name Set Affinity (Nsa) of Input Data Set of Procurement Process between Benchmarking Company and Best practice Company 2

		Best-practice Company (2)						
		Project Schedule Table (C01)	Project Budget (C02)	Project Process Method (C03)	Subcontractor List (C04)	Subcontractor Labors Activity (C05)	Financial Statement (C06)	Maximum of Row
Benchmarking Company	Purchasing Schedule (A01)	0.640	0.330	0.410	0.210	0.260	0.330	0.640
	Design Drawing (A02)	0.410	0.330	0.512	0.210	0.260	0.330	0.512
	Purchasing Specification (A03)	0.410	0.330	0.800	0.210	0.260	0.330	0.800
	Subcontractor List (A04)	0.210	0.210	0.210	0.800	0.210	0.210	0.800
Maximum of Column		0.640	0.330	0.800	0.800	0.260	0.330	
Sum of Max. of Row + Sum of Max. of Column								5.91
Nsa(P1(input set), P2(input set))								0.59

Table 4.5 Name Set Affinity (Nsa) of Output Data Set of Procurement Process between Benchmarking Company and Best practice Company 2

		Best-practice Company (2)										
		Procurement Schedule Table (C07)	Procurement Request Form (C08)	Inquiry Sheet (C09)	List of Certificated Subcontractors (C10)	Qualified Subcontractors List (C11)	Inquiry List (C12)	Subcontractor Quotation List (C13)	Comparison and Negotiation List (C14)	Detailed Prices List (C15)	Procurement Contract (C16)	Maximum of Row
Benchmarking Company	Procurement Schedule (A05)	0.800	0.260	0.260	0.210	0.210	0.260	0.260	0.260	0.210	0.260	0.800
	Procurement Request Form (A06)	0.260	0.800	0.410	0.210	0.210	0.410	0.410	0.512	0.210	0.330	0.800
	Inquiry Sheet (A07)	0.260	0.410	0.800	0.210	0.210	0.640	0.512	0.410	0.210	0.330	0.800
	Subcontractor Evaluation Table (A08)	0.210	0.210	0.210	0.410	0.512	0.210	0.210	0.210	0.410	0.210	0.512
	Qualified Subcontractors List (A09)	0.210	0.210	0.210	0.410	0.800	0.210	0.210	0.210	0.410	0.210	0.800
	Supplier Evaluation Table (A10)	0.210	0.210	0.210	0.410	0.512	0.210	0.210	0.210	0.410	0.210	0.512
	Quotation Request (A11)	0.260	0.410	0.512	0.210	0.210	0.512	0.640	0.410	0.210	0.330	0.640
	Price Evaluation Table (A12)	0.210	0.210	0.210	0.512	0.410	0.210	0.210	0.210	0.640	0.210	0.640
	Comparison and Negotiation List (A13)	0.260	0.512	0.410	0.210	0.210	0.410	0.410	0.800	0.210	0.330	0.800
	Detailed Prices Comparison List (A14)	0.210	0.210	0.210	0.512	0.410	0.210	0.210	0.210	0.640	0.210	0.640
	Procurement Contract (A15)	0.260	0.330	0.330	0.210	0.210	0.330	0.330	0.330	0.210	0.800	0.800
Maximum of Column		0.800	0.800	0.800	0.512	0.800	0.640	0.640	0.800	0.640	0.800	
Sum of Max. of Row + Sum of Max. of Column												14.98
Nsa(P1(output set), P2(output set))												0.71

Table 4.6 and Table 4.7 show the name set affinity of input data sets and output data sets of procurement process from benchmarking company (P_1), and best practice company 3 (P_4) of the case study, the $A(P_1.f_2, P_4.f_2)$ equals 0.560. Similarly, the output data set affinity parameters, $A(P_1.f_3, P_4.f_3)$ was also calculated that equals to 0.640. Accordingly, the $PISim(P_1, P_4) = 0.560 + 0.640 = 1.20$.

Table 4.6 Name Set Affinity (Nsa) of Input Data Set of Procurement Process between Benchmarking Company and Best practice Company 3

		Best-practice Company (3)			
		Project Budget (D01)	Project Schedule Table (D02)	Design Drawing (D03)	Maximum of Row
Benchmarking Company	Purchasing Schedule (A01)	0.330	0.640	0.410	0.640
	Design Drawing (A02)	0.330	0.410	0.800	0.800
	Purchasing Specification (A03)	0.330	0.410	0.512	0.512
	Subcontractor List (A04)	0.210	0.210	0.210	0.210
Maximum of Column		0.330	0.640	0.800	
Sum of Max. of Row + Sum of Max. of Column					3.93
Nsa(P1(input set), P2(input set))					0.56

Table 4.7 Name Set Affinity (Nsa) of Output Data Set of Procurement Process between Benchmarking Company and Best practice Company 3

		Best-practice Company (3)								
		Procurement Schedule Table (D04)	Procurement Budget (D05)	Contractor Evaluation Table (D06)	List of Certificated Subcontractors (D07)	Sub-working Fees (D08)	Notification List (D09)	Comparison and Negotiation List (D10)	Main Contract (D11)	Maximum of Row
Benchmarking Company	Procurement Schedule (A05)	0.800	0.330	0.210	0.210	0.210	0.260	0.260	0.260	0.800
	Procurement Request Form (A06)	0.260	0.260	0.210	0.210	0.210	0.330	0.512	0.330	0.512
	Inquiry Sheet (A07)	0.260	0.260	0.210	0.210	0.210	0.330	0.410	0.330	0.410
	Subcontractor Evaluation Table (A08)	0.210	0.210	0.640	0.410	0.410	0.210	0.210	0.210	0.640
	Qualified Subcontractors List (A09)	0.210	0.210	0.512	0.410	0.410	0.210	0.210	0.210	0.512
	Supplier Evaluation Table (A10)	0.210	0.210	0.640	0.410	0.410	0.210	0.210	0.210	0.640
	Quotation Request (A11)	0.260	0.260	0.210	0.210	0.210	0.330	0.410	0.330	0.410
	Price Evaluation Table (A12)	0.210	0.210	0.410	0.512	0.800	0.210	0.210	0.210	0.800
	Comparison and Negotiation List (A13)	0.260	0.260	0.210	0.210	0.210	0.330	0.800	0.330	0.800
	Detailed Prices Comparison List (A14)	0.210	0.210	0.410	0.512	0.640	0.210	0.210	0.210	0.640
	Procurement Contract (A15)	0.260	0.260	0.210	0.210	0.210	0.512	0.330	0.800	0.800
Maximum of Column		0.800	0.330	0.640	0.512	0.800	0.512	0.800	0.800	
Sum of Max. of Row + Sum of Max. of Column										12.16
Nsa(P1(output set), P2(output set))										0.64

4.2.2.2 Activity Similarity

The *PISim* function aims at expressing the similarity of input and output data sets; high process information similarity provides a macroscopic evaluation index which implies two processes have high potential for mergence due to their similar data features, and vice versa. However, a microcosmic view of form processing activities is necessary for advanced similarity analysis. Equation 4.11 expresses the activity similarity function denoted by $ASim(A_{hi}, A_{kj})$.

$$ASim(A_{ih}, A_{jk}) = NA(A_{ih}, A_{jk}) + A(AIN_{ih}, AIN_{jk}) + A(AOUT_{ih}, AOUT_{jk}) \ldots\ldots\ldots(4.11)$$

Where $0 \leqq ASim(A_{ih}, A_{jk}) \leqq 3$. A_{ih} is the name of hth activity of the *process i*; A_{jk} is the name of kth activity of the *process j*; $ASim(A_{ih}, A_{jk})$ is activity similarity of A_{ih} and A_{jk}; AIN_{ih} is the input set of A_{hi}; $AOUT_{jk}$ is the output set of A_{jk}.

In Equation 4.11, not only information similarity, expressed by $A(AIN_{ih}, AIN_{jk})$ and $A(AOUT_{ih}, AOUT_{jk}))$, but functional similarity, expressed by $NA(A_{ih}, A_{jk})$, is of concern. High activity similarity expresses the idea that two activities have a similarity in terms of work tasks.

4.2.2.3 Process Functional Similarity

Equation 4.12 summarizes all activity similarities of two processes that allows calculation of the Process Functional Similarity of P_i and P_j, denoted by $PFsim(P_i, P_j)$. Similar to *PISim* (P_i, P_j), the process functional similarity provides a macroscopic evaluation factor for determining whether two processes need to be integrated or not.

$$PFSim(P_i, P_j) = \frac{\sum_{h=1}^{m} \underset{k=1 \to n}{Max}(ASim(A_{ih}, A_{jk})) + \sum_{k=1}^{n} \underset{h=1 \to m}{Max}(ASim(A_{ih}, A_{jk}))}{m+n} \quad \forall A_{ih} \in P_i; \forall A_{jk} \in P_j \ldots\ldots\ldots(4.12)$$

where the $|P_i|$ and $|P_j|$ are numbers of activities of process-i and of process-j, and $0 \leqq PFSim(P_i, P_j) \leqq 3$.

The result of an analysis of procurement process between benchmarking company and best practice company 1 based on Equation 4.12 is shown in Table 4.8. The $FSim(P1,P2)$ equals 28.57/(10+10)= 1.43.

Table 4.8 Process Function Similarities (*FSim*) of Procurement Process between Benchmarking Company and Best practice Company 1

		Best-practice Company (1)										Maximum of Row
		Procurement Schedule	Procurement Verification	Subcontractor Information	Subcontractor Evaluation	Inquiries	Quotation	Comparison and Negotiation	Price Evaluation	Approval	Data Storage	
Benchmarking Company	Purchasing Schedule	1.012	0.510	0.335	0.335	0.385	0.385	0.385	0.385	0.385	0.125	1.01
	Purchasing Proposal	1.144	1.356	0.545	0.545	0.720	0.868	0.726	0.680	0.680	0.375	1.36
	Verification	1.000	1.737	0.545	0.545	1.011	1.031	0.805	0.766	0.760	0.406	1.74
	Collect Subcontractor Information	0.335	0.335	0.910	0.535	0.335	0.460	0.335	0.335	0.335	0.500	0.91
	Subcontractor Qualification	0.545	0.545	1.277	1.710	0.545	0.545	0.545	0.920	0.670	0.421	1.71
	Qualified Subcontractor Registration	0.545	0.545	1.460	1.149	0.545	0.920	0.545	0.545	0.545	0.500	1.46
	Supplier Evaluation	0.545	0.545	1.047	2.100	0.545	0.545	0.545	0.920	0.670	0.375	2.10
	Inquiries	0.620	0.746	0.645	0.645	1.733	0.894	1.052	0.876	0.790	0.447	1.73
	Comparison & Negotiation	0.595	0.670	0.796	0.745	0.921	0.760	1.313	0.753	0.714	0.421	1.31
	Approval	0.620	0.690	0.696	0.770	0.765	0.715	0.856	1.103	1.559	0.460	1.56
Maximum of Column		1.14	1.74	1.46	2.10	1.73	1.03	1.31	1.10	1.56	0.50	
Sum of Max. of Row + Sum of Max. of Column												28.57
Nsa(P1(input set), P2(input set))												1.43

Meanwhile, the result of an analysis of procurement process between benchmarking company and best practice company 2 based on Equation 4.12 is shown in Table 4.9. The $FSim(P1,P3)$ equals 28.10/(10+10)= 1.40.

Table 4.9 Process Function Similarities (*FSim*) of Procurement Process between Benchmarking Company and Best practice Company 2

		Best-practice Company (2)										Maximum of Row
		Procurement Schedule	Procurement Proposal	Procurement Verification	Subcontractor Information	Subcontractor Qualification	Inquiries	Quotation	Comparison and Negotiation	Approval	Data Storage	
Benchmarking Company	Purchasing Schedule	1.012	0.510	0.510	0.335	0.335	0.385	0.385	0.360	0.385	0.125	1.01
	Purchasing Proposal	1.266	1.482	1.187	0.545	0.545	0.720	0.755	0.683	0.663	0.475	1.48
	Verification	0.904	1.260	1.765	0.545	0.545	1.075	0.945	0.745	0.743	0.469	1.76
	Collect Subcontractor Information	0.335	0.335	0.335	1.140	0.535	0.335	0.460	0.486	0.335	0.500	1.14
	Subcontractor Qualification	0.545	0.545	0.545	1.335	1.652	0.545	0.545	0.645	0.771	0.421	1.65
	Qualified Subcontractor Registration	0.545	0.545	0.545	1.069	1.335	0.545	0.920	0.645	0.612	0.500	1.33
	Supplier Evaluation	0.545	0.545	0.545	0.944	1.422	0.545	0.545	0.645	0.737	0.375	1.42
	Inquiries	0.620	0.695	0.821	0.645	0.645	1.661	0.936	1.193	0.772	0.419	1.66
	Comparison & Negotiation	0.595	0.721	0.707	0.796	0.796	0.921	0.805	1.358	0.782	0.407	1.35
	Approval	0.620	0.690	0.753	0.696	0.821	0.765	0.727	0.705	1.713	0.446	1.71
Maximum of Column		1.26	1.48	1.76	1.33	1.65	1.66	0.94	1.35	1.71	0.50	
Sum of Max. of Row + Sum of Max. of Column												28.10
Nsa(P1(input set), P2(input set))												1.40

Finally, the result of an analysis of procurement process between benchmarking company and best practice company 3 based on Equation 4.12 is shown in Table 4.10. The $FSim(P1,P4)$ equals 24.12/(10+10)= 1.21.

Table 4.10 Process Function Similarities (*FSim*) of Procurement Process between Benchmarking Company and Best practice Company 3

		Best-practice Company (3)										Maximum of Row
		Procurement Planning	Contractor Evaluation	Certificated Subcontractor Publication	Sub-working Integration	Contractor Invitation	Quotation	Comparison and Negotiation	Price Evaluation	Approval	Data Storage	
Benchmarking Company	Purchasing Schedule	0.921	0.335	0.335	0.335	0.125	0.385	0.385	0.385	0.385	0.125	0.92
	Purchasing Proposal	1.071	0.722	0.545	0.665	0.335	0.812	0.721	0.511	0.680	0.426	1.07
	Verification	0.860	0.733	0.545	0.642	0.335	0.805	0.745	0.535	0.799	0.452	0.86
	Collect Subcontractor Information	0.335	0.535	0.890	0.637	0.500	0.460	0.335	0.335	0.335	0.500	0.89
	Subcontractor Qualification	0.545	1.350	0.944	0.745	0.765	0.696	0.847	0.710	0.670	0.421	1.35
	Qualified Subcontractor	0.545	0.972	1.550	0.745	0.535	1.020	0.745	0.460	0.545	0.375	1.55
	Supplier Evaluation	0.545	1.350	1.047	0.745	0.535	0.645	0.745	0.710	0.670	0.375	1.35
	Inquiries	0.620	0.695	0.696	1.265	0.335	0.630	0.770	0.435	0.805	0.419	1.27
	Comparison & Negotiation	0.595	0.670	0.796	0.910	0.486	0.765	1.510	0.630	0.705	0.407	1.51
	Approval	0.620	0.695	0.645	0.570	0.486	0.967	0.880	0.580	1.805	0.446	1.81
Maximum of Column		1.07	1.35	1.55	1.27	0.77	1.02	1.51	0.71	1.81	0.50	
Sum of Max. of Row + Sum of Max. of Column												24.12
Nsa(P1(input set), P2(input set))												1.21

4.3 Process Communication Index Analysis

To increase the success probability of BPR, the resistance value of each best-practice process needs to be evaluated. The process with the lowest resistance value should be the proper one to be adopted in benchmarking company in order to enhance the success probability of BPR. Based on this concept, the degree of communication ease from best-practice processes occurred in the organization structure of benchmarking company needs to be calculated. The total communication index (TCI) is proposed in this study to express the degree of communication ease of best-practice processes. The higher the communication index of best-practice process is, the smoother the best-practice process can be adapted in the benchmarking company. To evaluate the communication index, this study adopts trend model methodology to develop process communication index analysis.

Generally, the degree of communication ease from best-practice process occurred in benchmarking company is difficult to be evaluated. However, the trend model methodology proposed a proper method to evaluate the communication resistance of a process within organization. High process resistance implies low degree of communication ease of a process. Accordingly, the TCI of a process can be determined in accordance with the process resistance value evaluated by the trend model method. For this reason, following to trend model methodology, this study develops process communication index analysis as shown in Figure 4.19 to determine TCI showing the degree of communication ease.

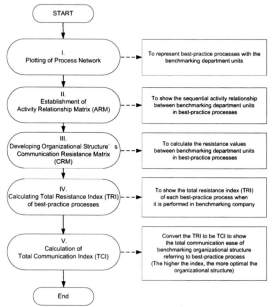

Figure 4.19 Process Communication Analyses

Figure 4.19 expresses the procedure of process communication analyses which includes five steps; namely, (1) plotting process network, (2) establishment of activity relationship matrix (ARM), (3) developing organizational structure's communication resistance matrix (CRM), (4) calculating total resistance index (TCI) of the best-practice process, and (5) Calculation of total communication index (TCI). With the application of the trend model, the process network that represents best-practice

66

processes with the benchmarking department units is created firstly. Subsequently, the activity relationship matrix (ARM) is developed to show the sequential activity relationship between benchmarking department units in best-practice processes. According to the result of the ARM, a communication resistance matrix (CRM) is generated to calculate the resistance values between benchmarking department units in best-practice processes. Then, the total resistance index (TRI) of best-practice processes when it is performed in benchmarking company is calculated accordingly. Finally, by converting the result of total resistance index (TRI), the total communication index (TCI) can be calculated to show the total communication ease of benchmarking organizational structure referring to best-practice processes. Based on the TCI resulted values from the aforementioned procedures, the best-practice process which can be executed most smoothly when it is adapted in benchmarking company can be determined.

4.3.1 Plotting of Process Network

The first step toward developing the process communication analysis is to create process network that describes the activities elements of a process in a logical hierarchy. Since the aim of benchmarking company is to get significant result based on best practice process, a replacement of department unit in best-practice process to be benchmarking department unit needs to be done.

The ARIS modeling language e-EPC diagram is also applied to produce visual representations of the abstract process model. Figure 4.20, Figure 4.21 and Figure 4.22 show the e-EPC diagram of procurement process model from benchmarking company referring to best practice company 1, best practice company 2 and best practice company 3.

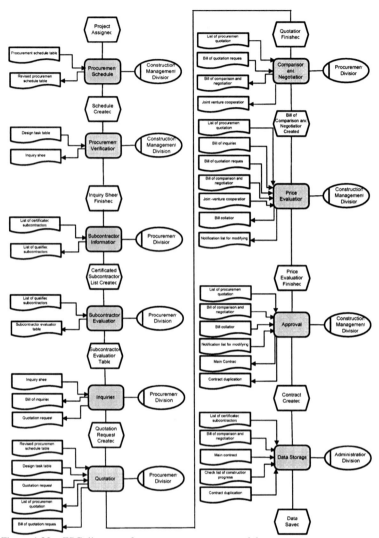

Figure 4.20 e-EPC diagram of procurement process model
[Benchmarking Company referring to Best practice company 1 Process]

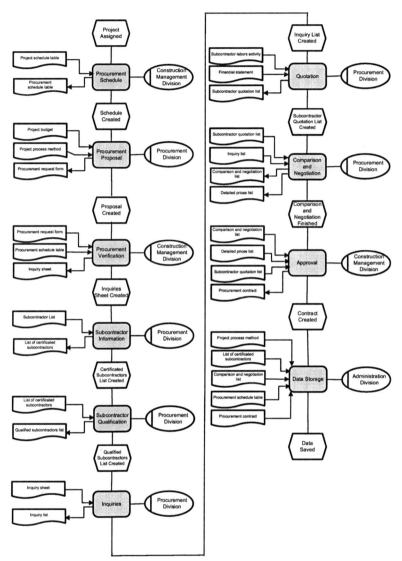

Figure 4.21 e-EPC diagram of procurement process model
[Benchmarking Company referring to Best practice company 2 Process]

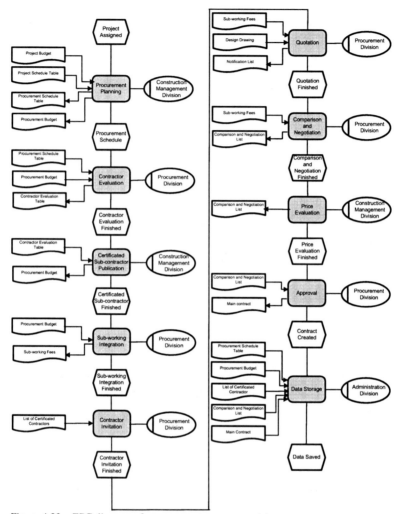

Figure 4.22 e-EPC diagram of procurement process model
[Benchmarking Company referring to Best practice company 3 Process]

4.3.2 Establishment of Activity Relationship Matrix (ARM)

Referring to the previous step which is plotting process network, Activity Relationship Matrix (ARM) is created. The establishment of the ARM in this study is to show the sequential relationship between the departments and activities in procurement process. Based on e-EPC diagram of procurement process model which is shown in Figure 4.20, Figure 4.21 and Figure 4.22, the numbers of handovers between the departments can be found and presented in a matrix. Table 4.11, Table 4.12 and Table 4.13 present Activity Relationship Matrix (ARM) of benchmarking company referring to best practice company 1 procurement process, best practice company 2 procurement process and best practice company 3 procurement process.

Table 4.11 Activity Relationship Matrix (ARM) Benchmarking Company referring to Best practice Company 1 Procurement Process

		Succeeding Activity Unit		
		Construction Management Division	Procurement Division	Administration Division
Preceding Activity Unit	Construction Management Division	2	1	1
	Procurement Division	1	4	0
	Administration Division	0	0	0

Table 4.12 Activity Relationship Matrix (ARM) Benchmarking Company referring to Best practice Company 2 Procurement Process

		Succeeding Activity Unit		
		Construction Management Division	Procurement Division	Administration Division
Preceding Activity Unit	Construction Management Division	2	2	1
	Procurement Division	2	4	0
	Administration Division	0	0	0

Table 4.13 Activity Relationship Matrix (ARM) Benchmarking Company referring to Best practice Company 3 Procurement Process

		Succeeding Activity Unit		
		Construction Management Division	Procurement Division	Administration Division
Preceding Activity Unit	Construction Management Division	0	3	0
	Procurement Division	2	3	1
	Administration Division	0	0	0

4.3.3 Developing Organizational Structure's Communication Resistance Matrix

According to the result of the ARM, a communication resistance matrix (CRM), which displays the resistance between benchmarking department unit in best-practice process, is developed. Communication resistance matrix (CRM) is composed of communication resistance value between two departments of different layers within organizational structure that is represented by *Ki*. Resistance coefficient *Ki* is the basic variable used to represent the degree of communication ease in the project. To estimate the weight of *Ki*, a questionnaire was sent and 10 responses were valid. Figure 4.23 shows communication resistance of benchmarking organization structure.

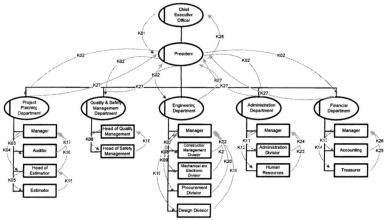

Figure 4.23 Communication Resistance of Benchmarking Organization Structure

There are two steps are involved in the creation of a Communication Resistance Matrix (CRM).

Step1. Create Resistance Coefficient Evaluation of Questionnaire.

The purpose for this survey is to investigate the relative weight of the communication resistance value *(Ki)* between two members at different levels of the organizational structure. According to the respective working contracts and their roles in the construction sequence, the relative positions of departments in the project's organizational structure are determined. Departments in different layers may have different degrees of communication ease with the related parties in higher or lower layers

Resistance coefficient *(Ki)*, a one-way communication index, is used to evaluate the ease of communication for solving disputes, conflicts, or coordination problems between related parties in different layers. The basic assumptions of this survey are as follows: (1) only mutual reliance within the operation and each party's niche position in the project's organizational structure are the factors considered for evaluating the Ki value. (2) only members in adjacent layers have a contractual relationship; for example, President has direct contractual relationship with his/her higher layer CEO and lower layer Head of Quality Management, and Head of Safety Management has no direct contractual relationship with President of the company.

Since this study focuses on procurement process, only the resistance coefficient from departments that are related to the procurement process which are engineering department and administration department need to be calculated. Figure 4.24 shows communication resistance of procurement process in benchmarking organization structure.

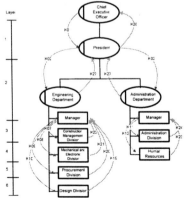

Figure 4.24 Communication Resistance of Procurement Process in
Benchmarking Organization Structure

Using Figure 4.24 as an example, the definitions of the Ki values are as follows. $K01$ = the resistance of the CEO to the members of the first layer; $K02$ = the resistance of members of the first layer to those in the second layer; $K27$ = the resistance of members of the second layer to those in the first layer; and $K28$ = the resistance of members of the first layer to the CEO. Definitions in the rest of the K_i were determined in the same manner.

Survey Questions. The survey questions are as follows.

1. Based on your personal experience and understanding of construction projects, please rank the Ki values.

2. All of the resistance values Ki shown in Figure 4.23 were determined on a pairwise comparison basis. For example, in Table 4.14, if you think the communication resistance of members of the first layer to those of the second layer ($K2$) is more difficult than the communication resistance of the CEO to the members of the first layer ($K1$), the "difficult" column is checked.

Ki	Very Easy		Easy		Equal		Difficult		Very Difficult	value
	5	4	3	2	1	1/2	1/3	1/4	1/5	
K01	-	-	-	-	-	-	v	-	-	K02
K01	-	-	-	-	-	v	-	-	-	K07
K01	-	-	-	-	v	-	-	-	-	K08
K01	-	-	v	-	-	-	-	-	-	K09
K01	-	-	-	-	-	v	-	-	-	K10
K01	-	-	v	-	-	-	-	-	-	K11
K01	-	-	-	-	v	-	-	-	-	K12
K01	-	-	-	-	-	-	v	-	-	K19
K01	-	-	-	-	-	v	-	-	-	K20
K01	-	-	-	v	-	-	-	-	-	K21
K01	-	-	v	-	-	-	-	-	-	K22
K01	-	-	-	-	v	-	-	-	-	K23
K01	-	-	-	-	-	-	v	-	-	K24
K01	-	-	-	-	-	-	v	-	-	K27
K01	-	-	-	-	-	-	v	-	-	K28
K02	-	-	-	-	-	v	-	-	-	K07
K02	-	-	-	v	-	-	-	-	-	K08
K02	-	-	-	-	-	-	-	v	-	K09
K02	-	-	-	-	-	-	-	-	v	K10
K02	-	-	-	v	-	v	-	-	-	K11
K02	-	-	v	-	-	-	-	-	-	K12
K02	-	-	-	-	v	-	-	-	-	K19
K02	-	-	-	-	-	v	-	-	-	K20
K02	-	-	-	-	-	-	-	v	-	K21
K02	-	-	-	-	-	-	-	v	-	K22
K02	-	-	v	-	-	-	-	-	-	K23
K02	-	-	-	-	v	-	-	-	-	K24
K02	-	-	-	-	-	-	v	-	-	K27
K02	-	-	-	v	-	-	-	-	-	K28
K07	-	-	-	-	-	-	-	-	-	K08
K07	-	-	-	-	-	v	-	-	-	K09
K07	-	-	-	-	-	-	-	v	-	K10
K07	-	-	-	v	-	-	-	-	-	K11
K07	-	v	-	v	-	-	-	-	-	K12
K07	-	-	-	v	-	-	-	-	-	K19
K07	-	v	-	-	-	-	-	-	-	K20
K07	-	v	-	-	-	-	-	-	-	K21
K07	-	-	-	v	-	-	-	-	-	K22
K07	-	-	-	-	-	-	v	-	-	K23
K07	-	-	-	-	-	v	-	-	-	K24
K07	-	v	-	-	-	-	-	-	-	K27
K07	-	-	v	-	-	-	-	-	-	K28
K08	-	-	-	v	-	-	-	-	-	K09
K08	-	-	-	-	-	-	-	-	-	K10
K08	-	-	-	-	v	-	-	-	-	K11
K08	-	-	-	v	-	-	-	-	-	K12
K08	-	-	-	v	-	-	-	-	-	K19
K08	-	-	-	v	-	-	-	-	-	K20
K08	-	-	-	-	-	-	v	-	-	K21
K08	-	-	-	-	-	v	-	-	-	K22
K08	-	-	-	-	-	-	v	-	-	K23
K08	-	-	-	-	-	v	-	-	-	K24
K08	-	-	-	-	-	v	-	-	-	K27
K08	-	-	-	-	v	-	-	-	-	K28
K09	-	-	-	v	-	-	-	-	-	K10
K09	-	-	v	-	-	-	-	-	-	K11
K09	-	-	v	-	-	-	-	-	-	K12
K09	-	-	-	-	-	-	-	v	-	K19
K09	-	-	-	-	-	-	v	-	-	K20
K09	-	-	-	v	-	-	-	-	-	K21
K09	-	v	-	-	-	-	-	-	-	K22
K09	-	v	-	-	-	-	-	-	-	K23
K09	-	v	-	-	-	-	-	-	-	K24
K09	-	-	-	-	v	-	-	-	-	K27
K09	-	-	-	-	v	-	-	-	-	K28
K10	-	-	-	-	v	-	-	-	-	K11
K10	-	-	-	-	-	-	v	-	-	K12
K10	-	-	-	-	-	-	v	-	-	K19
K10	-	-	-	-	-	v	-	-	-	K20
K10	-	-	-	-	v	-	-	-	-	K21
K10	-	-	-	v	-	-	-	-	-	K22
K10	-	-	v	-	-	-	-	-	-	K23
K10	-	-	v	-	-	-	-	-	-	K24
K10	-	-	v	-	-	-	-	-	-	K27
K10	-	v	-	-	-	-	-	-	-	K28
K11	-	v	-	-	-	-	-	-	-	K12
K11	-	-	-	-	v	-	-	-	-	K19
K11	-	-	-	-	v	-	-	-	-	K20
K11	-	-	-	-	-	-	v	-	-	K21
K11	-	-	-	-	-	-	v	-	-	K22
K11	-	-	-	-	v	-	-	-	-	K23
K11	-	-	-	-	-	-	-	v	-	K24
K11	-	-	-	-	v	-	-	-	-	K27
K11	-	-	-	-	v	-	-	-	-	K28
K12	-	-	-	v	-	-	-	-	-	K19
K12	-	-	v	-	-	-	-	-	-	K20
K12	-	-	v	-	-	-	-	-	-	K21
K12	-	-	-	-	-	-	-	v	-	K22
K12	-	-	-	-	-	-	-	v	-	K23
K12	-	-	-	-	-	-	-	v	-	K24
K12	-	-	-	-	-	-	v	-	-	K27
K12	-	-	-	-	v	-	-	-	-	K28
K19	-	-	-	v	-	-	-	-	-	K20
K19	-	-	-	-	v	-	-	-	-	K21
K19	-	-	v	-	-	-	-	-	-	K22
K19	-	-	v	-	-	-	-	-	-	K23
K19	-	-	-	-	v	-	-	-	-	K24
K19	-	-	-	-	-	v	-	-	-	K27
K19	-	-	-	-	-	v	-	-	-	K28
K20	-	-	-	-	v	-	-	-	-	K21
K20	-	-	-	v	-	-	-	-	-	K22
K20	-	-	-	v	-	-	-	-	-	K23
K20	-	-	-	-	-	-	v	-	-	K24
K20	-	-	-	-	-	v	-	-	-	K27
K20	-	-	-	-	-	v	-	-	-	K28
K21	-	-	-	-	-	-	-	v	-	K22
K21	-	-	-	-	-	-	-	v	-	K23
K21	-	v	-	-	-	-	-	-	-	K24
K21	-	v	-	-	-	-	-	-	-	K27
K21	-	-	-	-	-	v	-	-	-	K28
K22	-	-	-	-	v	-	-	-	-	K23
K22	-	-	-	-	-	-	v	-	-	K24
K22	-	-	-	-	-	v	-	-	-	K27
K22	-	-	-	-	-	-	v	-	-	K28
K23	-	v	-	-	-	-	-	-	-	K24
K23	-	-	-	v	-	-	-	-	-	K27
K23	-	-	-	v	-	-	-	-	-	K28
K24	-	-	-	-	-	-	-	v	-	K27
K24	-	-	-	-	-	-	v	-	-	K28
K27	-	-	-	-	-	-	v	-	-	K28

Table 4.14 Pairwise Comparison

76

After transforming the linguistic items in Table 4.14 to the represented values, all questionnaires underwent a consistency test to ensure that they were logically correct. The data, suitable for general procurement processes, are used as the basis for a quantitative evaluation of the optimal organizational structure of benchmarking company referring to best practice companies procurement process. Table 4.15 shows inconsistent matrix, table 4.16 shows normalized matrix, table 4.17 shows the Consistency Ratio (CR) calculation and Table 4.18 shows overall priorities of Ki

Table 4.15 Inconsistent Matrix of Pairwise Comparison

	K01	K02	K07	K08	K09	K10	K11	K12	K19	K20	K21	K22	K23	K24	K27	K28
K01	1.00	3.00	2.00	3.00	3.00	2.00	0.33	0.33	0.33	0.33	0.33	0.33	0.50	0.50	0.50	3.00
K02	0.33	1.00	0.33	0.33	0.50	0.33	0.33	0.33	0.33	0.33	0.50	0.33	0.25	0.33	0.50	2.00
K07	0.50	3.00	1.00	1.00	0.50	0.50	0.33	0.25	0.50	0.50	1.00	1.00	0.25	0.25	1.00	3.00
K08	0.33	3.00	1.00	1.00	0.33	0.50	0.25	0.25	0.50	0.50	1.00	0.50	0.25	0.25	1.00	1.00
K09	0.33	2.00	2.00	3.00	1.00	0.50	0.33	0.33	0.50	1.00	2.00	2.00	0.33	0.33	2.00	2.00
K10	0.50	3.00	2.00	2.00	2.00	1.00	0.33	0.33	1.00	2.00	2.00	2.00	0.33	0.33	2.00	2.00
K11	3.00	3.00	3.00	4.00	3.00	3.00	1.00	0.33	4.00	4.00	4.00	4.00	0.50	1.00	4.00	4.00
K12	3.00	3.00	4.00	4.00	3.00	3.00	3.00	1.00	4.00	4.00	4.00	4.00	1.00	1.00	4.00	4.00
K19	3.00	3.00	2.00	2.00	2.00	1.00	0.25	0.25	1.00	0.50	0.50	0.50	0.50	0.50	1.00	1.00
K20	3.00	3.00	2.00	2.00	1.00	0.50	0.25	0.25	2.00	1.00	1.00	1.00	0.33	0.33	1.00	1.00
K21	3.00	2.00	1.00	1.00	0.50	0.50	0.25	0.25	2.00	1.00	1.00	1.00	0.33	0.33	1.00	1.00
K22	3.00	3.00	1.00	2.00	0.50	0.50	0.25	0.25	2.00	1.00	1.00	1.00	0.33	0.33	1.00	1.00
K23	2.00	4.00	4.00	4.00	3.00	3.00	2.00	1.00	2.00	3.00	3.00	3.00	1.00	0.50	2.00	2.00
K24	2.00	3.00	4.00	4.00	3.00	3.00	1.00	1.00	2.00	3.00	3.00	3.00	2.00	1.00	2.00	2.00
K27	2.00	2.00	1.00	1.00	0.50	0.50	0.25	0.25	1.00	1.00	1.00	1.00	0.50	0.50	1.00	0.33
K28	0.33	0.50	0.33	1.00	0.50	0.50	0.25	0.25	1.00	1.00	1.00	1.00	0.50	0.50	3.00	1.00
Column Total	27.33	41.50	30.67	35.33	24.33	20.33	10.42	6.67	24.17	24.17	26.33	25.67	8.92	8.00	27.00	30.33

Table 4.16 Normalized Matrix, Row Sums

	Normalized Matrix																Row Sums	Average
	K01	K02	K07	K08	K09	K10	K11	K12	K19	K20	K21	K22	K23	K24	K27	K28		
K01	0.037	0.072	0.065	0.085	0.123	0.098	0.032	0.050	0.014	0.014	0.013	0.013	0.056	0.063	0.019	0.099	0.852	0.053
K02	0.012	0.024	0.011	0.009	0.021	0.016	0.032	0.050	0.014	0.014	0.019	0.013	0.028	0.042	0.019	0.066	0.389	0.024
K07	0.018	0.072	0.033	0.028	0.021	0.025	0.032	0.037	0.021	0.021	0.038	0.039	0.028	0.031	0.037	0.099	0.580	0.036
K08	0.012	0.072	0.033	0.028	0.014	0.025	0.024	0.037	0.021	0.021	0.038	0.019	0.028	0.031	0.037	0.033	0.473	0.030
K09	0.012	0.048	0.065	0.085	0.041	0.025	0.032	0.050	0.021	0.041	0.076	0.078	0.037	0.042	0.074	0.066	0.793	0.050
K10	0.018	0.072	0.065	0.057	0.082	0.049	0.032	0.050	0.041	0.083	0.076	0.078	0.037	0.042	0.074	0.066	0.923	0.058
K11	0.110	0.072	0.098	0.113	0.123	0.148	0.096	0.050	0.165	0.165	0.152	0.156	0.056	0.125	0.148	0.132	1.910	0.119
K12	0.110	0.072	0.130	0.113	0.123	0.148	0.288	0.150	0.165	0.165	0.152	0.156	0.112	0.125	0.148	0.132	2.290	0.143
K19	0.110	0.072	0.065	0.057	0.082	0.049	0.024	0.037	0.041	0.021	0.019	0.019	0.056	0.063	0.037	0.033	0.786	0.049
K20	0.110	0.072	0.065	0.057	0.041	0.025	0.024	0.037	0.083	0.041	0.038	0.039	0.037	0.042	0.037	0.033	0.781	0.049
K21	0.110	0.048	0.033	0.028	0.021	0.025	0.024	0.037	0.083	0.041	0.038	0.039	0.037	0.042	0.037	0.033	0.676	0.042
K22	0.110	0.072	0.033	0.057	0.021	0.025	0.024	0.037	0.083	0.041	0.038	0.039	0.037	0.042	0.037	0.033	0.728	0.045
K23	0.073	0.096	0.130	0.113	0.123	0.148	0.192	0.150	0.083	0.124	0.114	0.117	0.112	0.063	0.074	0.066	1.778	0.111
K24	0.073	0.072	0.130	0.113	0.123	0.148	0.096	0.150	0.083	0.124	0.114	0.117	0.224	0.125	0.074	0.066	1.833	0.115
K27	0.073	0.048	0.033	0.028	0.021	0.025	0.024	0.037	0.041	0.041	0.038	0.039	0.056	0.063	0.037	0.011	0.615	0.038
K28	0.012	0.012	0.011	0.028	0.021	0.025	0.024	0.037	0.041	0.041	0.038	0.039	0.056	0.063	0.111	0.033	0.592	0.037

Table 4.17 Totaling the Entries

	K01 (0.053)	K02 (0.024)	K03 (0.036)	K04 (0.030)	K05 (0.050)	K06 (0.058)	K07 (0.119)	K08 (0.143)	K09 (0.049)	K10 (0.049)	K11 (0.042)	K12 (0.045)	K13 (0.111)	K14 (0.115)	K15 (0.038)	K16 (0.037)	Row Sums
K01	0.053	0.072	0.072	0.090	0.150	0.116	0.040	0.048	0.016	0.016	0.014	0.015	0.056	0.058	0.019	0.111	0.945
K02	0.018	0.024	0.012	0.010	0.025	0.019	0.040	0.048	0.016	0.016	0.021	0.015	0.028	0.038	0.019	0.074	0.423
K07	0.027	0.072	0.036	0.030	0.025	0.029	0.040	0.036	0.025	0.025	0.042	0.045	0.028	0.029	0.038	0.111	0.635
K08	0.018	0.072	0.036	0.030	0.017	0.029	0.030	0.036	0.025	0.025	0.042	0.023	0.028	0.029	0.038	0.037	0.512
K09	0.018	0.048	0.072	0.090	0.050	0.029	0.040	0.048	0.025	0.049	0.084	0.090	0.037	0.038	0.076	0.074	0.867
K10	0.027	0.072	0.072	0.060	0.100	0.058	0.040	0.048	0.049	0.098	0.084	0.090	0.037	0.038	0.076	0.074	1.022
K11	0.159	0.072	0.108	0.120	0.150	0.174	0.119	0.048	0.196	0.196	0.168	0.180	0.056	0.115	0.152	0.148	2.160
K12	0.159	0.072	0.144	0.120	0.150	0.174	0.357	0.143	0.196	0.196	0.168	0.180	0.111	0.115	0.152	0.148	2.585
K19	0.159	0.072	0.072	0.060	0.100	0.058	0.030	0.036	0.049	0.025	0.021	0.023	0.056	0.058	0.038	0.037	0.892
K20	0.159	0.072	0.072	0.060	0.050	0.029	0.030	0.036	0.098	0.049	0.042	0.045	0.037	0.038	0.038	0.037	0.892
K21	0.159	0.048	0.036	0.030	0.025	0.029	0.030	0.036	0.098	0.049	0.042	0.045	0.037	0.038	0.038	0.037	0.777
K22	0.159	0.072	0.036	0.060	0.025	0.029	0.030	0.036	0.098	0.049	0.042	0.045	0.037	0.038	0.038	0.037	0.831
K23	0.106	0.096	0.144	0.120	0.150	0.174	0.238	0.143	0.098	0.147	0.126	0.135	0.111	0.058	0.076	0.074	1.996
K24	0.106	0.072	0.144	0.120	0.150	0.174	0.119	0.143	0.098	0.147	0.126	0.135	0.222	0.115	0.076	0.074	2.021
K27	0.106	0.048	0.036	0.030	0.025	0.029	0.030	0.036	0.049	0.049	0.042	0.045	0.056	0.058	0.038	0.012	0.688
K28	0.018	0.012	0.012	0.030	0.025	0.029	0.030	0.036	0.049	0.049	0.042	0.045	0.056	0.058	0.114	0.037	0.640

Determining λ max:

$$
\begin{bmatrix} 0.945 \\ 0.423 \\ 0.635 \\ 0.512 \\ 0.867 \\ 1.022 \\ 2.160 \\ 2.585 \\ 0.892 \\ 0.892 \\ 0.777 \\ 0.831 \\ 1.996 \\ 2.021 \\ 0.688 \\ 0.640 \end{bmatrix}
\div
\begin{bmatrix} 0.053 \\ 0.024 \\ 0.036 \\ 0.030 \\ 0.050 \\ 0.058 \\ 0.119 \\ 0.143 \\ 0.049 \\ 0.049 \\ 0.042 \\ 0.045 \\ 0.111 \\ 0.115 \\ 0.038 \\ 0.037 \end{bmatrix}
=
\begin{bmatrix} 17.75 \\ 17.39 \\ 17.54 \\ 17.30 \\ 17.49 \\ 17.72 \\ 18.10 \\ 18.06 \\ 18.15 \\ 18.27 \\ 18.40 \\ 18.26 \\ 17.95 \\ 17.64 \\ 17.89 \\ 17.29 \end{bmatrix}
$$

Average of λ max:

$$\frac{17.75 + 17.39 + 17.54 + 17.30 + 17.49 + 17.72 + 18.10 + 18.06 + 18.15 + 18.27 + 18.40 + 18.26 + 17.95 + 17.64 + 17.89 + 17.29}{16} = 17.83$$

Determining Consistency Ratio (where CI = 1.61):

$$\frac{\lambda \max - n}{n - 1} = \frac{17.83 - 16}{16 - 1} = 0.122$$

Consistency Ratio:

= 0.122 / 1.61

= 0.08

= 8%, which indicates good consistency.

Table 4.18 overall priorities of *Ki*

Value	Overall Priorities
K01	0.053
K02	0.024
K07	0.036
K08	0.030
K09	0.050
K10	0.058
K11	0.119
K12	0.143
K19	0.049
K20	0.049
K21	0.042
K22	0.045
K23	0.111
K24	0.115
K27	0.038
K28	0.037

Step 2. Developing Communication Resistance Matrix.

The communication of procurement process is assumed to be transferred through the layer structures as shown in Figure 4.23; thus, messages passed from Manager of Engineering Department to Manager of Administration Department should go through President and President passes messages received from Manager of Engineering Department to Manager of Administration Department. Based on this assumption, resistance should be accumulated; thus, resistance from Manager of Department Engineering to Manager of Administration Engineering is the sum of resistance values from Manager of Engineering Department to President and from President to Manager of Administration Department. Accordingly, the values for communication resistance among members of the project team could be calculated in the same manner and tabulated in matrix form. Table 4.19, Table 4.20 and Table 4.21 show the communication resistance matrix benchmarking company referring to best practice company 1 procurement process, best practice company 2 procurement process and best practice company 3 procurement process.

Table 4.19 Communication Resistance Matrix (CRM) Benchmarking Company referring to Best-practice Company 1 Procurement Process

		Succeeding Activity Unit		
		Construction Management Division	Procurement Division	Administration Division
Preceding Activity Unit	Construction Management Division	0	K20+K07	K24+K27+ K02+K07
	Procurement Division	K22+K09	0	K24+K27+ K02+K09
	Administration Division	K22+K27+ K02+K11	K20+K27+ K02+K11	0

Table 4.20 Communication Resistance Matrix (CRM) Benchmarking Company referring to Best-practice Company 2 Procurement Process

		Succeeding Activity Unit		
		Construction Management Division	Procurement Division	Administration Division
Preceding Activity Unit	Construction Management Division	0	K20+K07	K24+K27+ K02+K07
	Procurement Division	K22+K09	0	K24+K27+ K02+K09
	Administration Division	K22+K27+ K02+K11	K20+K27+ K02+K11	0

Table 4.21 Communication Resistance Matrix (CRM) Benchmarking Company referring to Best-practice Company 3 Procurement Process

		Succeeding Activity Unit		
		Construction Management Division	Procurement Division	Administration Division
Preceding Activity Unit	Construction Management Division	0	K20+K07	K24+K27+ K02+K07
	Procurement Division	K22+K09	0	K24+K27+ K02+K09
	Administration Division	K22+K27+ K02+K11	K20+K27+ K02+K11	0

4.3.4 Calculating Total Resistance of Best-practice Processes

After activity relationship matrix (ARM) and communication relationship matrix (CRM) have been developed, total resistance index (TRI) is calculated accordingly to present the total resistance of best-practice processes when it is performed in benchmarking organizational structure. The one with the smallest value is determined to be the least conflict best-practice process when it is adapted in the benchmarking company.

Multiplying all of the values of the Activity Relationship Matrix (ARM) shown in Table 4.11, Table 4.12 and Table 4.13 by the corresponding values of the Communication Resistance Matrix (CRM) shown in Table 4.18, Table 4.19 and Table 4.20 produce the total resistance matrix (Tmn) for that particular organizational structure. Equation 4.13 shows the Total Resistance Matrix of organizational Structure (Tmn), meanwhile Equation 4.14 shows the Total Resistance Index (TRI) which is the sum of all of the components of the Tmn.

Total resistance matrix of organizational structure

Tmn = Activity Relationship Matrix **.** Communication Resistance Matrix.........(4.13)
where : (**.**) = product symbol; and m,n = members of project organizational structure.

Total Resistance Index (TRI) = Σ Tmn...(4.14)

The calculated result of the total resistance matrix for the example is shown in Table 4.22, Table 4.23 and Table 4.24. Meanwhile the calculated result of Total Resistance Index (TRI) is shown in Table 4.25

Table 4.22 Total Resistance Calculation of Benchmarking Company referring to Best-practice Company 1 Procurement Process

Preceding Activity Unit	Succeeding Activity Unit		
	Construction Management Division	Procurement Division	Administration Division
Construction Management Division	2	1	1
Procurement Division	1	4	0
Administration Division	0	0	0

Preceding Activity Unit	Succeeding Activity Unit		
	Construction Management Division	Procurement Division	Administration Division
Construction Management Division	0	K20+K07	K24+K27+K02+K07
Procurement Division	K22+K09	0	K24+K27+K02+K09
Administration Division	K22+K27+K02+K11	K20+K27+K02+K11	0

Preceding Activity Unit	Succeeding Activity Unit		
	Construction Management Division	Procurement Division	Administration Division
Construction Management Division	0	K20+K07	K24+K27+K02+K07
Procurement Division	K22+K09	0	0
Administration Division	0	0	0

Table 4.23 Total Resistance Calculation of Benchmarking Company referring to Best-practice Company 2 Procurement Process

Preceding Activity Unit	Succeeding Activity Unit		
	Construction Management Division	Procurement Division	Administration Division
Construction Management Division	2	2	1
Procurement Division	2	4	0
Administration Division	0	0	0

Preceding Activity Unit	Succeeding Activity Unit		
	Construction Management Division	Procurement Division	Administration Division
Construction Management Division	0	K20+K07	K24+K27+K02+K07
Procurement Division	K22+K09	0	K24+K27+K02+K09
Administration Division	K22+K27+K02+K11	K20+K27+K02+K11	0

Preceding Activity Unit	Succeeding Activity Unit		
	Construction Management Division	Procurement Division	Administration Division
Construction Management Division	0	2(K20+K07)	K24+K27+K02+K07
Procurement Division	2(K22+K09)	0	0
Administration Division	0	0	0

Table 4.24 Total Resistance Calculation of Benchmarking Company referring to Best-practice Company 3 Procurement Process

Preceding Activity Unit	Succeeding Activity Unit		
	Construction Management Division	Procurement Division	Administration Division
Construction Management Division	0	3	0
Procurement Division	2	3	1
Administration Division	0	0	0

Preceding Activity Unit	Succeeding Activity Unit		
	Construction Management Division	Procurement Division	Administration Division
Construction Management Division	0	K20+K07	K24+K27+K02+K07
Procurement Division	K22+K09	0	K24+K27+K02+K09
Administration Division	K22+K27+K02+K11	K20+K27+K02+K11	0

Preceding Activity Unit	Succeeding Activity Unit		
	Construction Management Division	Procurement Division	Administration Division
Construction Management Division	0	3(K20+K07)	0
Procurement Division	2(K22+K09)	0	K24+K27+K02+K09
Administration Division	0	0	0

Table 4.25 Total Resistance Index of Benchmarking Company

Company	Total Resistance Index (TRI)
Best-practice Company [1]	0.393
Best-practice Company [2]	0.573
Best-practice Company [3]	0.656

4.3.5 Calculation of Total Communication Index (TCI)

After Total Resistance Index (TRI) for each best practice process to be performed in benchmarking company has been calculated, total communication index (TCI) can be converted accordingly from the TRI to present the total degree of communication ease. The higher the communication index of best-practice process, the smoother the best-practice process can be adapted in the benchmarking company, so that the feasibility for implementing the best-practice processes might be enhanced. Based on the TRI's concept, TCI can be presented mathematically as shown in Equation 4.15.

Total Communication Index (TCI) = $1 - TRI*$ ....(4.15)

Where: $0 \leq TCI \leq 1$; TRI* represents normalized total resistance index (TRI).

The calculated result of the Total Communication Index (TCI) is shown in Figure 4.26

Table 4.26 Total Communication Index of Benchmarking Company

	Organization View		
	TRI	TRI*	TCI
Best-practice Company [1]	0.393	0.599	0.401
Best-practice Company [2]	0.573	0.873	0.127
Best-practice Company [3]	0.656	1.000	0.000
Maximun Value	0	0	1

Based on the evaluation results, the best practice process which can be executed most smoothly in benchmarking company organization structure is best practice company 1, because of its highest TCI value. Comparing the procurement process of best-practice company 1 with other best-practice companies, it can be seen that best-practice company 1 has fewer conflicts in communication and is more efficient in coordination when it is implemented in the benchmarking company.

83

4.4 Process Adaptability Calculation

Based on the results of process similarity analysis and process communication index analysis of best practice processes, the adaptability index (AI) for each best-practice process can be calculated. Process adaptability calculation is developed to calculate an adaptability index that represents the acceptance degree of each best practice process for benchmarking company. The higher the AI is, the more suitable the process to be adapted in benchmarking company.

To calculate the AI, two primary tasks are necessary in this step. Firstly, the relative weight of PI Sim, PF Sim and TCI need to be calculated; secondly, since the scales of PI Sim, PF Sim, and TCI values are different, these three indexes can not be summarized to AI value directly. Therefore, the normalizations need to be applied to set these three indexes within the range 0 to 1.

To quantify the relative weight, the AHP method is adopted for each index as shown in Figure 4.25. Therefore two pairwise comparison needs to be created to quantify the relative weight between process similarity and process communication index; and the relative weight between PI Sim and PF Sim. Table 4.27 shows the pairwise comparison between process similarity and process communication index. Table 4.28 shows the pairwise comparison between PI Sim and PF Sim.

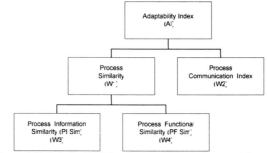

Figure 4.25 Relative Weight of Adaptability Index (AI) Hierarchy

Table 4.27 Adaptability Index (AI) Pairwise Comparison between
Process Similarity and Process Communication Index

Ki	Not Important		Less Important		Equal		Important		Very Important	value
	5	4	3	2	1	1/2	1/3	1/4	1/5	
Process Similarity	-	-	-	-	-	-	V	-	-	Process Communication Index

Table 4.28 Adaptability Index (AI) Pairwise Comparison between
PI Sim and PF Sim

Ki	Not Important		Less Important		Equal		Important		Very Important	value
	5	4	3	2	1	1/2	1/3	1/4	1/5	
PI Sim	-	-	-	-	-	-	-	V	-	PF Sim

Survey Questions. The survey questions are as follows.

1. Based on your personal experience and understanding of construction projects, please rank process similarity, process communication index, PI Sim and PF Sim values.

2. All of the relative weight between process similarity and process communication; and the relative weight between PI Sim and PF Sim. were determined on a pairwise comparison basis. For example, in Table 4.27, if you think the function objects of process (PF Sim) is more important than the data objects of process (PI Sim), the "important" column is checked.

After transforming the linguistic items in Table 4.27 and Table 4.28 to the represented values, the relative weight for each factor can be calculated. Table 4.29 shows inconsistent matrix of process similarity and process communication index. Table 4.30 shows relative weight of process similarity and process communication index. Table 4.31 shows inconsistent matrix of PI Sim and PF Sim. Table 4.32 shows relative weight of PI Sim and PF Sim.

Table 4.29 Inconsistent Matrix of Process Similarity and
Process Communication Index

	Process Similarity	Process Communication Index
Process Similarity	1.000	0.667
Process Communication Index	1.500	1.000
Column Total	2.500	1.667

Table 4.30 Relative Weight of Process Similarity and
Process Communication Index

	Process Similarity	Process Communication Index	Row Sums	Relative Weight
Process Similarity	0.400	0.400	0.800	0.40
Process Communication Index	0.600	0.600	1.200	0.60

Table 4.31 Inconsistent Matrix of PI Sim and PF Sim

	Process Information Similarity	Process Functional Similarity
Process Information Similarity	1.000	0.435
Process Functional Similarity	2.300	1.000
Column Total	3.300	1.435

Table 4.32 Relative Weight of PI Sim and PF Sim

	Process Information Similarity	Process Functional Similarity	Row Sums	Relative Weight
Process Information Similarity	0.303	0.303	0.606	0.30
Process Functional Similarity	0.697	0.697	1.394	0.70

Thus, based on the calculation, the relative weight result for Adaptability Index

(AI) hierarchy is shown in Figure 4.26.

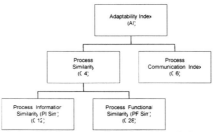

Figure 4.26 Relative Weight Results for Adaptability Index (AI) Hierarchy

86

After the relative weights for PI Sim, PF Sim, and TCI have been calculated, the range value (Xi) for each index (i) needs to be set from 0 to 1 ($0 \leq Xi \leq 1$). For this reason, PI Sim, and PF Sim need to be normalized since these indexes have different range scales ($0 \leq$ PI Sim ≤ 2 ; $0 \leq$ PF Sim ≤ 3) . Table 4.33 shows Process Information Similarity (PI SIM) Normalization and Table 4.34 shows Process Functional Similarity Normalization (PF SIM).

Table 4.33 Process Information Similarity (PI SIM) Normalization

	Data View	
	PI Sim	Normalized PI Sim
Best-practice Company [1]	1.220	0.610
Best-practice Company [2]	1.300	0.650
Best-practice Company [3]	1.200	0.600
Maximun Value	2	1

Table 4.34 Process Functional Similarity (PF SIM) Normalization

	Function View	
	PF Sim	Normalized PF Sim
Best-practice Company [1]	1.430	0.323
Best-practice Company [2]	1.400	0.316
Best-practice Company [3]	1.210	0.273
Maximun Value	3	1

Following to the relative weights that have been quantified and range value are set to be zero to one, the adaptability index of each best-practice company can finally be calculated. The one with the highest value is determined to be the most suitable best practice process to be adapted in benchmarking company.

Equation 4.16 shows Adaptability Index (AI) of a process

Adaptability Index (AI) = W1.PIij+W2.PFij+W3.TCij....................................(4.16)

Where : W1 = Relative weight of PI Sim; W2 = Relative weight of PF Sim; W3 = Relative weight of TCI; PIij = Process Information Similarity value of a process; PFij = Process Functional Similarity value of a process; TCij = Total Communication Index of a process. (i) = the name of best-practice companies, (j) = the index values.

The result for the Adaptability Index (AI) of procurement process is shown in Table 4.35

Table 4.35 Adaptability Index (AI) of Procurement Process

	Process Similarity (0.4)		Process Communication Index (0.6)	Adaptability Index [AI]	Adaptability Index [AI]
	Information Similarities (0.12)	Functional Similarities (0.28)			
Best-practice Company [1]	0.610	0.323	0.401	0.40	40%
Best-practice Company [2]	0.650	0.316	0.127	0.24	24%
Best-practice Company [3]	0.600	0.273	0.000	0.15	15%

Based on the analysis result of process similarity and process communication index of procurement process, the best practice company which has the highest value to be adapted in benchmarking company organization structure is best practice company 1. Comparing best practice company 1 with the other best practice companies for process similarity, it can be seen that the gap value between best practice company 1, best practice company 2 and best practice company 3 is not large due to resemblance of data objects and function objects of procurement process. Nevertheless, in terms of process communication index, benchmarking company 1 has a fewer conflicts in communication and is more efficient for coordination. Thus, the most suitable process from the best-practice companies to be adapted in benchmarking company is best-practice company 1 after considered two factors; namely process similarity and process communication index.

Chapter 5

Conclusion and Recommendations

This chapter presents research objectives, summarizes findings and assessments that extend objectives were achieved. Research contributions and lessons learned from BOPR experiments are also addressed. In conclusion, this chapter discusses recommendations as to future prospects for the BOPR.

5.1 Review of Objectives

The primary purpose of this study is to develop a systematic enterprise analysis method to assist project team in determining the most suitable process from best practice processes to be performed in benchmarking company's. The BOPR method is subsequently developed to assist the benchmarking company to achieve the primary goal. The sub-objectives achieved to facilitate BOPR implementation are included (1) Following the business process reengineering philosophy in applying benchmarking philosophy to identify the performance of best practice process and to intrigue the process to be adopted; (2) Developing process similarity analysis to confirm the corresponding semantic relationships of each compared data object pair and function object pair between best-practice processes and benchmarking process; (3) Developing process communication index analysis to evaluate the degree of communication ease of best practice processes to be performed in benchmarking company organization; (4) Developing process adaptability analysis to create an Adaptability Index (AI) that represents the acceptance degrees of each best practice company's process to be performed in benchmarking company.

5.2 Summary

The Benchmarking Oriented Process Reengineering (BOPR) is proposed to facilitate the project team in determining which the most suitable process from the best practice companies to be adapted in benchmarking company. The most suitable process is the process which has similar characteristics with benchmarking process and can be executed smoothly when it will be performed in benchmarking company. Therefore, four phases must be incorporated in the BOPR, including (1) business process modeling, (2) process similarities analysis, (3) process communication index analysis, and (4) process adaptability calculation.

Applying ARIS modeling language e-EPC diagram and textual process model to develop business process modeling

In the process of establishment BOPR method, a business process modeling needs to be schemed preliminarily and the relationship between the data, view and organization view of one process need to be created clearly. By applying ARIS modeling language eEPC diagram and textual process model, a process model that shows the relationship between data, function and organization view can be represented. ARIS modeling tool is used to create the original process model based on the collected data; the textual process model is used to create data-oriented process model which is necessary for the process similarity analysis in the next phase.

Applying semantic similarity to evaluate process similarities analysis

Since one process consists of several activities and numerous data, the similarity between data and function from benchmarking and best practice companies are essential for BOPR method. Process similarity analysis is developed to evaluate the data and activities similarity between the best-practice processes and benchmarking process. Process similarity is summarized by process information similarity (PI Sim) and process functional similarity (PF Sim). The higher the PI Sim is, the more similar input or output data the two processes have; similarly, the higher the PF Sim is, the more similar activities the two processes have. For the purpose, semantic similarity philosophy is applied to evaluate the PI Sim and PF Sim.

Applying trend model concept to analyze the communication ease of the process

Since the selected best-practice process will be implemented by the benchmarking company, the degree of communication ease of best-practice process occurred in benchmarking company needs to be calculated. The higher the communication index of best-practice process, the smoother the best-practice process can be adapted in the benchmarking company.

Generally, the degree of communication ease from best-practice process occurred in benchmarking company is difficult to be evaluated. However, the trend model methodology proposed a proper method to evaluate the communication resistance of a process within organization. High process resistance implies low degree of communication ease of a process. Accordingly, the communication index of a process can be calculated in accordance with the analyzed process resistance value in the trend model method.

BOPR method is proposed to facilitate project team to determine the most suitable process

The BOPR method was proposed to facilitate project team in determining which best-practice processes is the most suitable to be adapted in benchmarking company. By using BOPR method, an adaptability index (AI) is created to show the acceptance degree of each best-practice process for benchmarking company. The higher the AI is, the more suitable the process to be adapted in benchmarking company.

Adaptability index is composed two factors which are process similarity and process communication resistance, that are represented by Process Information Similarity (PI Sim), Process Functional Similarity (PF Sim) and Total Communication Index (TCI). Based on the AI results, a project team can determine the most suitable process from best practice companies.

BOPR feasibility is validated through a case study of procurement process from one benchmarking company and three companies which are represented as best practice companies in construction industry. Analysis results of case study have not only demonstrated the appropriateness of applying each phases of BOPR method but, also validated BOPR method concept feasibility.

5.3 Conclusions

Following to BPR philosophy, a benchmarking philosophy is applied to develop the benchmarking oriented process reengineering (BOPR) method. The BOPR method provides a systematic analysis method to determine the most suitable process from best practice companies to be performed in benchmarking company. By using the BOPR method, the most suitable best-practice process which has similar characteristics with benchmarking process and can be executed smoothly in benchmarking organization structure can be determined. Conclusions based on research observations and experiences are followed:

Benchmarking can be applied to BPR for process redesign

By applying benchmarking philosophy to redesign a process, an unnecessary task in BPR can be reduced such as the trial and error time in redesigning a process. Consequently, the duration of BPR can be shortened since the benchmarking provides a best-to-learn paradigm from the selected best-practice companies in redesigning a process.

Process similarity can be evaluated by applying semantic similarity analysis.

This research focuses on procurement process within construction industry. Determining commonalities between two processes is a key to identify the similarity between benchmarking company and best practice companies procurement process. Since a process is composed of several activities and data entities, an analysis regarding the activities and data entities needs to be evaluated. This study applies semantic similarities analysis to evaluate process similarity.

Process similarity can be calculated by clustering all data entities and activity names of the procurement process. Furthermore, data and activity hierarchies which describe the semantic distance between either two data entities or two activity names are created. Based on semantic hierarchy, affinity functions are developed to calculate data similarities and activity names similarities. The Process Information Similarity (PI Sim), Activity Similarity (A Sim), and Process Functional Similarity (PF Sim) of two processes can be calculated to identify the commonalities between two processes.

The degree of communication ease of process can be evaluated by applying trend model concept.

A process is categorized as an effective and efficient process if during the execution of the process has no conflicts in communication. In order to increase the success probability of BPR, the communication resistance of the process needs to be evaluated. This study applies trend method methodology to develop the process communication index analysis. By following the trend model methodology, the degree of communication ease can be calculated based on communication resistance of a process.

Summarizing process similarity and process communication index, the AI can be calculated to express the acceptance degrees of best-practice processes.

In the BOPR method, the adaptability index (AI) is calculated to represent the acceptance degrees of best-practice processes for benchmarking company. Based on the research assumption, the most suitable process has most similar characteristics with benchmarking company and can be executed smoothly when it will be adapted in benchmarking organization structure, AI can be calculated by summarizing process similarity and process communication index. Process similarity expresses the similar characteristics between best-practice processes and benchmarking process; the process communication index evaluates the degree of communication ease from best-practice processes when it will be adapted in benchmarking company

BOPR facilitates benchmarking project team to analyze and determine best practice companies.

In the BOPR method, it consists of four phases that each phase presents a systematic procedure in analyzing a process and generates a detailed analysis of process. Furthermore, BOPR method facilitates a benchmarking project team in determining and implementing the best practice company process in benchmarking company.

5.4 Research Contributions

The contributions of this research include

(1) This research develops a process similarities analysis which is summarized by PI Sim and PF Sim to evaluate the commonalities between two processes

(2) This research develops process communication index analysis to evaluate the degree of communication index of best-practice process occurred in

benchmarking organization structure.

(3) This research creates an Adaptability Index (AI) that represents an acceptance degree for determining the most suitable process to be implemented in benchmarking company.

(4) This research successfully develops Benchmarking Oriented Process Reengineering (BOPR) method to propose a systematic analysis method for determining the most suitable process from best-practice company so that the success of BPR might be increased

5.5 Discussions

Implementation of Benchmarking Oriented Process Reengineering (BOPR)

This study pioneers a systematic analysis method to determine the best-to-learn process from best practice companies for the benchmarking organization. BOPR is developed to facilitate project team to determine which best-practice is most suitable to be adapted in benchmarking company. Due to the numerous working interfaces, complicated networks and diversified team members of construction projects, coordination of documents and efficiency among members of each construction company may differs with the other construction company.

BOPR implementation issue: The information regarding three best practice companies that have been conducted shows that the similarity of data view and function view between best practice companies has a slightly difference. Since this issue has occurred, a further modification of information gathering process may be necessary. Although in analysis of data view, function view and organization view of the process shows the similarity between best practice companies has a higher difference, but an additional of relationships and classifications of best practice companies process is still needed to be generated. Accordingly the project team can append newly created relationships and classifications of the created models to make it practicable.

5.6 Recommendations

Three recommendations are suggested for future research on this subject.

Defining a semantic data thesaurus will be a critical element in analyzing the similarities between two processes in the construction industry.

A data thesaurus provides a common linguistic framework essential to sharing and integrating information between two independent units. It gathers and provides the semantic relationships and the corresponding affinity values between data distributed over different units. Furthermore, the data thesaurus provides the common framework that allows examination or assessment of relationships and similarities between data items. Moreover, the data thesaurus is essential to further analysis of activity and functional process in the construction industry. As creating a uniform data thesaurus will demand a significant investment in time and other resources, a suggestion would be using the data dictionary of the construction industry as a starting point and reference for development. This has already prepared reference that gathers relevant construction industry terminology and assigns corresponding data attributes.

Defining the resistance index is the key factor for realizing the organizational structure.

This study proposes a quantitative evaluation method for determining the most suitable procurement process to be applied in benchmarking organization structure. In future applications of this method for determining the most suitable process to be applied in organization structure, the resistance index is the key factor for realizing the organizational structure. Thus, it is suggested that some modifications be made to the resistance index based on the characteristics and working conditions of the construction project.

Considering cost factor in process adaptability calculation can predict the expenditures of benchmarking company in adapting best-practice process.

The most suitable process in this study is the process which has the most similar characteristics with benchmarking process and the process can be executed smoothly in benchmarking company. To achieve this goal, BOPR method creates Adaptability Index (AI) which is the acceptance degree for benchmarking company to determine which the most suitable process is. AI considers two factors; namely, process similarity and process communication. In future applications of this method for determining the most suitable process, it is suggested that a cost factor could be a significant factor to be considered in calculating the AI of best-practice processes. By considering the cost factor, the benchmarking company can predict the expenditures to adapt each of best-practice company's process.

BIBLIOGRAPHY

Andrea, M., Egenhofer Max J. (2003). "Determining Semantic Similarity among Entity Classes from Different Ontologies." IEEE Transactions onn Knowledge and Data Engineering, 15(2), 442-456.

Bennigson, L. A. ~1971!. "TREND: New management information from networks." *Proc., 3rd Int. Congress on Project Planning by Network* Techniques, 44–60.

Camp, Robert C. "Benchmarking". ASQ.1989. New York

Castano, S. and De Antonellis, V. (1995a). "Reengineering processes in public administrations". In Proc. *of OO-ERJ95- Int. Conf. on the Object-Oriented and Entity-Relationship Modelling,.* 282-295, Gold Coast, Australia.

Castano, S. and De Antonellis, V. (1995b). "Reference conceptual architectures for re-engineering information systems." International Journal of Cooperative Information Systems, 49(2&3):213-235.

Castano, S. and De Antonellis, V. (1997) "A multi-perspective framework for the analysis of legacy information systems." In Proc. of CAiSE'97 - Int. Conf. on Advanced Information Systems Engineering, 117-130, Barcelona, Spain.

Castano, S., De Antonellis,V. (1998). "A framework for expressing semantic relationships between multiple information systems for cooperation", Information Systems 23(3/4), pp.253-277.

De Antonellis, V. and Zonta, B. (1990). "A Disciplined Approach to Offine Analysis.

" IEEE Transactions on Software Engineering, 16(8), 822-828.

Furnas, G.W., Landauer, I.K., Gomez, L.M., and Dumais, S.T. (1987). "The vocabulary problem in human-system communication." *Communications of the ACM*, 30(11), 964-971

Galbraith, J.R.(1973). "Designing Complex Organizations". Addison-Wesley Publishing Company.

James T. O'Connor and Steven J. Miller, "Constructability Programs: Method for Assessment and Benchmarking." Journal of Performance of Constructed Facilities. 8(1), 1994, pp.46-64.

Li, Y., Bandar, Z.A., Mclean, D. (2003). "An approach for measuring semantic similarity between words using multiple information sources," IEEE Transactions on Knowledge and Data Engineering, 15(4), 871 - 882.

M. Hammer, Reengineering work: Don't automate, obliterate, Harvard Business Review, (1990) 104-112.

M. Hammer, J. Champy, Reengineering the corporation-A manifesto for business revolution, Harper Collins, New York, 1993.

Min-Yuan, Cheng., Cheng-Wei., Su. and Horng-Yuh, You, "Optimal Project Organizational Structure for Construction Management". Journal of Construction Engineering and   Management 129(1), 2003, pp.70-79.

Min-Yuan. Cheng, M.H. Tsai, "Reengineering of Construction Management Process." Journal of Construction Engineering and Management. 129(1), 2003, pp.105-114.

M.H. Tsai, (2007). Reengineering of Cross-organization Process for Design-Build Projects. Ph.D. Dissertation, National Taiwan University of Science and Technology.

Planning, organizing and managing Benchmarking activities: User's guide." (1992). Am. Productivity and Quality Ctr., Houston, Tex.

Poulson, B., 1996, Process benchmarking in retail financial services. Management Services, June, 12–14.

Ricardo R. Ramirez, Luis Fernando C. Alarcon and Peter Knights, "Benchmarking System for Evaluating Management Practice in the Construction Industry." Journal of Management in Construction. 20(3), 2004, pp.110-117.

Sang-Hoon Lee, Stephen R. Thomas, and Richard L. Tucker. "Web-Based Benchmarking System for the Construction Industry." Journal of Construction Engineering and Management. 131(7), 2005, pp.790-798.

Scheer, A. W., (2000), ARIS: Business Process Modeling, (Berlin: Springer).

Sciore, E., Siege, M., and Rosenthal, A. (1994). "Using semantic values to facilitate interoperability among heterogeneous information systems". ACM Tbnsactions on Database Systems, 19(2), 254-290.

Sheth, A. and Kashyap, V. (1992). "So Far (Schematically) Yet So Near (Semantically)." Proc. IFIP WG2.6 Database Semantic Conf. Interoperable Database Systems. DS-5, 283-312.

Spendolini, Michael J. "The Benchmarking Book". Amacom. 1992. New York.

Taiichi, O. (1990). Toyota production System: Beyond large scale production. Productivity Press, Cambridge, Mass.

Thamhain, H. J. (1991). "Benchmarking of project management systems: How to measure yourself against the best." Seminar Proc., Project Management Institute, Dallas, Tex., 471-476.

Tenner, A. R. and Detoro, I.J., 1997, *Process Redesign: The Implementation Guide for Managers* (Reading, MA: Addison Wesley)

Watson, G. H. (1993). Strategic benchmarking: How to rate your company's performance against the world's best. John Wiley and Sons, Inc.,, London, England.

Y, C. Juan and C. Ou-Yang., 2004, Systematic approach for the gap analysis of business process. International Journal of Production Research Vol.42, No. 7, 1325-1364.

Zairi, M. (1992). "Competitive Benchmarking: An executive guide," Technical Communications Ltd., Letchworth, U.K.

Printed in the United States
116907LV00001B/88/P

9 783639 004519